HOMO SAPIENS, A PROBLEMATIC SPECIES

An Essay in Philosophical Anthropology

Mia Gosselin

University Press of America,® Inc.
Lanham · Boulder · New York · Toronto · Plymouth, UK

4501 Forbes Boulevard
Suite 200
Lanham, Maryland 20706
UPA Acquisitions Department (301) 459-3366

Unit A, Whitacre Mews, 26-34 Stannary Street,
London SE11 4AB, United Kingdom

British Library Cataloging in Publication Information Available

Library of Congress Control Number: 2014954311
ISBN: 978-0-7618-6519-3 (paperback : alk. paper)
eISBN: 978-0-7618-6520-9

Contents

Preface

The path of an academic career, at least in Belgium, but no doubt also in other countries, is not always predictable. At the Free University of Brussels the task of teaching philosophical anthropology was assigned tot me in my younger years, though it was not my speciality. Probably it was expected of me that I would teach this discipline in the spirit of existentialism, an offshoot of German idealism which was very popular on the European continent. However, I had studied the work of Nietzsche, who like his *maître à penser*, Schopenhauer, was an anti-Hegelian. Being convinced that science based on empirical data offers the best guarantee for understanding nature and ourselves, even if, or rather because, the results of science can always be critically assessed and replaced by new ones if necessary, I rejected idealism and metaphysical explanations. A new type of philosophical anthropology seemed necessary.

Perhaps as a reaction to existentialism, its idealistic roots and its strong literary components, I specialised in the drier matter of epistemology and the foundation of science. I had received a thorough education in the history of philosophy, for which I am still grateful, and studying the growth over long periods of time of the Western conception of man and of his place in nature, I became aware of the contrast between this general conception as expressed in the work of philosophers and what we can learn about our species in different scientific disciplines.

The genealogy of Western anthropological views is the subject of the first volume of my book. In the second volume I study the scientific conception of the human species, how science pictured over time the nature of life and the origin of species in general and that of our own species in particular. I also try to understand what science says about, *inter alia,* what is specific to our species that makes us so destructive for our environment, why we are aggressive to our congeners, and many other features of our kind, most of them related to its problematic character.

It is not the place in the preface to this first volume to elaborate on the themes of the second volume, but the link between the first and the second one must already become clear. On the one hand Westerners have a specific view of the world that is anthropocentric, whereas on the other hand science considers our species to be only one amongst all others and studies it as a natural phenomenon. It is this unresolved contradiction that I want to expose.

Westerners have acquired a dominant position in the world and especially in world economy, from which follows its political power. It is in Europe that modern science developed and that, through technology making use of its results, the industrial revolutions took place, which in their turn gave rise to the welfare society of today. In many countries of the third world Western lifestyle was and is seen as identical with progress. It is not always understood that at its base is a fundamentally anthropocentric world view in which man is considered as a reality apart from and even opposed to nature. The loss of the harmony between man and nature has important negative consequences. My aim in the first volume is to examine how this world view came about over long stretches of time and to lay bare its roots.

Here follows a short overview of its content.

- Part I. The general conception in prehistoric and ancient cultures was that the world was one (monophysism): nature, the Gods, the ghosts, all plants and animals, men, rocks and rivers formed one whole. First I try to grasp the psychology behind primitive thought (Chapter 1). Secondly, I try to understand the primitive conception of what it is to be a human individual and to analyse the logic of this type of thought. The latter is completely different from the logic Aristotle introduced in the West that was generally adopted for practical and scientific matters in Antiquity, but also in the Middle Ages and thereafter. (Logic is to be understood in its general sense of what is accepted by a cultural community as valid reasoning.) (Chapter 2). Even today our logic in daily life and in science is basically Aristotelian. Aristotle rejected the logic of his master Plato and stressed the advantages of his own theories on the subject. However, transcendent metaphysics claims that Aristotelian logic is not suitable for philosophy and reverts to more primitive ways of reasoning. (I think the fact that in European culture, save in religious and metaphysical matters, Aristotelian logic was used, explains in an important, but not exclusive way, why modern science developed in Europe). (Chapter 3).

- Part II. The first Greek philosophers saw man as a natural phenomenon, an integral part of the rest of reality. To them the major question was how out of the original chaos the ordered world, the cosmos, had arisen. Absorbed by scientific and practical matters, religious and moral questions were not central. They were famous in their own time, but a fundamentally different philosophical approach to reality replaced their views. Man was in this new idealistic conception exalted as a spiritual being, the only creature possessing a soul reaching out for a reality behind reality. Atomism, a successor of the philosophy of nature, combating superstition, never became as popular as philosophies that promised under one form or another life after death (Chapter 4).

The religious source of inspiration, Judaeo-Christianity, is of course all-important. Man is created by God to whom he owes his existence and he is given by him the assignment to reign over nature. This separate creation of mankind means that we are willed by God, chosen to exist. We are a necessary part of the divine plan and have a mission; we must contribute to the realisation of the Kingdom of God on Earth. At the moment of the conception of each individual God breathes life in him or her and thus we are endowed with a soul determining what kind of person we become. We have free will and are judged by our deeds, but also by our intentions.

Christian religion, in order to become well-established in the Hellenistic world, had to give its rather simple doctrine an intellectual structure that could appeal to more educated people and more powerful members of society. The spiritual leaders of early Christianity saw the dangers for their faith contained in Greek philosophy and the latter was only acceptable as, like it was termed later on, the "servant maid of theology." It seemed to the Church Fathers that the best choice for their purpose was Platonism and neo-Platonism. The eternal soul, its purity and life after death were for the believer what really mattered. The body and its urges had to be kept in check; the material world was unimportant in the light of the promise of salvation for those who were pure of heart. Aristotelian logic (called dialectics in the Middle Ages) should not be applied to matters of faith, because it could lead to conclusions that were in contradiction with the main dogmas, such as the Holy Trinity. It is only in the thirteenth century that the philosophy of Aristotle, after a long struggle and in a Christian version, was accepted by the authorities as an alternative to Platonism. The eternal soul and life after death remained the central issue.

Though at the end of the Middle Ages and in the Renaissance the picture of our solar system profoundly changed and though the excellence of man was no longer only due to the grace of God, but to his own creative force, the picture of man remained the same on a more fundamental level. The earth now turned around the sun rather than the other way round, but man was still the crown on the work of God (Chapter 5).

In Modern Times we can witness a profound contrast between rationalists and empiricists: if all knowledge comes from God and the concepts stem from him, we are the only rational beings; however, if we acquire knowledge through our senses, so do the animals and there is a certain gradual continuity between the species, between the animals most closely related to us and ourselves. It is rather difficult to know what the real religious convictions of these philosophers were, because in their time, the seventeenth and eighteenth century, they could not freely speak their mind. Descartes was always cautious not to offend the religious authorities. He insisted that the soul and free will existed, because otherwise there was no foundation for morality in the prospect of reward after death for living the good life. Animals had no soul; they were complicated and well-functioning machines. Hume on the contrary, who was a sceptic where biblical truth was concerned, expressed his conviction that neither the soul (an ego or self) nor free will existed. To him the continuity between animals and men was an evidence. We have to take into account that these philosophers, besides having different philosophical views, lived in different countries and in different periods (Chapter 6).

We could easily be tempted to believe that with the evermore rapid progress made by science everything would change at the end of the eighteenth and nineteenth century where the picture of man and his relation with the rest of nature is concerned, but if we read Hegel and Marx, respectively an idealist and Deist and a materialist and atheist, we realise that both have a profoundly anthropocentric view and that for both all that matters to man is man. Hegel humanises nature by thinking it, Marx humanises it by transforming it by labour. Reality is a dynamic process that of necessity must lead to an end that can be compared to what for the Christian is the establishment of God's Kingdom on Earth at the end of times. For Hegel it is the realisation of the World-spirit, for Marx the establishment of a society without classes. Reality must still accomplish itself through the thoughts and the actions of men, but that the final goal will be reached cannot be doubted. In both cases, the motor of the process is man, who has a mission in this world (Chapter 7).

In spite of a number of precursors, the first and to my knowledge only philosophers to break explicitly and radically with this millenary tradition of anthropocentrism and the belief the soul is a substance (i.e. can exists in its own right) are two atheists, Schopenhauer and Nietzsche. Schopenhauer insists that man *is* his body and by this fact is part of nature that is the expression of the Will, *das Ding an sich*, (reality as it is *in se*). Kant, who invented the concept, believed that reality as it is in itself could not be known by man, but Schopenhauer claimed that it could be intuited speculatively. In Nature there is only one law, the struggle for life, which man also obeys, leaving him bereft of free will. The world is not good, rational or divine. It is a vale of tears. When we witness the cruelty of nature, we must accept that we are not different. Influenced by Eastern thought, he sees as a solace universal compassion. Nietzsche rejects Schopenhauer's metaphysics, his pessimism and universal compassion. Nevertheless, he has inherited many of his fundamental ideas. Thus he believes man is part of nature; we are a species amongst species. He adopts evolutionism and either we are going to evolve into a new kind of creature he does not define, save as utterly different from the species we are now, a creature he calls "Overman" or "Superman," or we become a superlative of what we are now, and we become "the last man," who in his eyes is despicable. He is an outspoken antihumanist and never tires of exposing anthropocentrism and of showing it is a farce. He denies free will and rejects subjectivism. Body and mind are one. The role of consciousness and self-consciousness is exaggerated by idealist philosophers. Too much of them can even be negative, because by too much reflection and self-reflection we are prevented from spontaneously following our instincts, which he is the first philosopher to value positively. General moral values are to be replaced by the values each individual creates for himself.

Nietzsche became very famous, but he was not understood. The Nazis claimed him for their racist cause, he was wrongly considered a vitalist (vitalists see Life as a creative force, a substance existing in its own right), the existentialists saw in him a forerunner and some religious philosophers even denied he was a genuine atheist. His influence on the picture we have of ourselves and our place in nature was minor (Chapter 8).

Again, in the twentieth century we find the same old claims: reality is ontologically layered and man is unique by being at once a material and a spiritual being, by relating the material world to a transcendent reality. Man is necessary, he has a mission. Philosophical anthropology was an invention of the heirs of German idealism. They wanted to give an answer to the problem posed to metaphysics by Darwinism. Man could not be simply a species like all other species, he had to have unique qualities that explained the necessity of his existence and his role in this world. The common denominator of these anthropologies is clear: man is a unique kind of reality.

Next I treat of Teilhard de Chardin, who was a priest and a theologian, but also an archaeological anthropologist, quite famous for his scientific work. He courageously tried to reconcile his Christian faith and evolutionism. The Roman Catholic Church did not appreciate his views and forbade him to pursue this endeavour. He believed that species had evolved (not being a biologist, he was wrong on several points about how evolution took place). Behind evolution there was a divine plan, he assumed. Everything tended to an ultimate outcome, to God, or *Point Omega*, in which the circle was closed, an ever returning idea we can find in the philosophy of Plotinus.

The last section of this final chapter is devoted to Sartre, because I want to show that though he is an atheist, by inscribing himself in an old tradition and by declaring himself a humanist, he reiterates many themes of the Christian conception of reality under one form or another.

His starting point is the three H's: Hegel, Husserl, Heidegger. Reality is the product of our subjectivity; we are an ego or, but not an empirical one, determined by a genetic component and by a history of past experiences. It is an abstract ego that cannot be described in terms of qualities, but nevertheless lends us individuality. I can know who I am after I made all my choices at each moment of my existence. This is the explanation of his rejection of essentialism and of his famous dictum 'existence precedes essence.' We are not born with a specific soul instilled in us by God, that will determine at each moment our behaviour. This ego we are, and this is crucial, is on the contrary absolute freedom. If you ask if the choices I made were good or bad, the answer is that the only justification of my acts is that they are *my* acts. Sartre would have liked to be *causa sui*, his own cause and he thinks in our inner selves we all want this. As a child he experienced a crisis of identity: he was not necessary, he could as well not have been. Therefore, a recurring theme is this contingency he could not accept. Christians believe that they exist because God has willed it, or at least allowed it. Sartre lacks this faith and wants to become necessary.

Man is what is important to man. Like Hegel, Sartre is a Manichean: unreflective Nature is what he abhors most. A tree is there outside, but it is there as an *en soi*, an entity that is merely as it is; it is not there *pour soi*, for itself, because it cannot think itself. It needs a human being thinking: 'it is a tree,' to become one. It is man who creates reality; without human subjectivity there would only be a second order reality, things as they are without reflection.

Sartre declares himself a humanist. Just like each human being by his or her deeds determines what he or she will have been at the end of his or her existence, each human being contributes to the essence of humanity. For all his atheism, Sartre remains a Christian, though one without God, because, like all Westerners, he is an heir of the most profound convictions of his culture. Free will is the cornerstone of Christian morality. Like Hegel, Sartre made out of freedom the hallmark of humanity, but, like so many others, he was blind to the fact it is an illusion. Ask the mother in the African Sahel desert, cradling her dying child in her arms, whether she can make choices and is free. Even if she assumes her destiny (or if you prefer her situation and that of her child) it is hypocrisy to say that all that matters is her unique way to assume it. If nobody helps her, she has no choice but to die with her child. If indeed she cannot choose, according to Sartre she is not a human being (Chapter 9).

It is of course not because of the predominance of Western culture and its peculiar nature that we are a problematic species. The reasons for it are to be found in characteristics common to humans in general. Yet the combination of these characteristics with the ideas and messages Westerners consciously or unconsciously spread, with their unwavering belief of progress through technology, enhances the destructive processes that are ongoing.

I want to thank my colleague and friend Joris Duytschaever who has read the manuscript and who corrected the language. The eventually remaining mistakes are wholly my responsibility and due to the fact that I am not a native speaker of English. Another dear friend, Martin De Beukelaer, made a number of very useful remarks, for which I am grateful.

I am the author of the translations of German and French cited texts except if otherwise mentioned. Occasionally I have used square brackets in these texts to add an explanative note. When citing a philosopher if I left out for brevity's sake some lines in a paragraph I signal this by using (…).

Antwerp, ……

Mia Gosselin.

Part I.

Before Philosophy

Chapter 1. The psychology of mythical thought

The first true philosophers were Greeks of the sixth century B.C. It was their vocation to explain how the natural order that we see around us came into being, constructing rational models, hypotheses, *which could be discussed*, starting from their sensorial experience. Instead of asking *why* things are as they are like in speculative thought, common in all cultures before them, they asked *how* they came about. They were philosophers and scientists at the same time, preoccupied by the general principles governing nature and not by ethical questions. In the following century quite other, even opposite, types of philosophy developed. This is not our subject here, but rather "what was there before philosophy?" The answer is myths and in a more developed form religion instead of theoretical knowledge of nature. Myths result from the capacity of humans to explain what is inexplicable using their imagination.

It is commonly admitted that religion is universal. Different cultures offer mythical religious world views that determine the conception we have of ourselves and of fellow human beings and therefore they are crucial for our subject, namely the origin of our view of man. The question why religion is universal has been answered in different ways. Scientists who were staunch believers were convinced that it is essential to human nature, as animals show no sign of even considering if the world makes sense and if they, as part of it, make sense; they are only occupied with the bare necessities of life. This explanation is about as accurate as the answer of the doctor in one of the plays of Molière to the question why opium induces sleep: by its *virtus dormitiva*, by its force to make sleepy.

Other scientists have asked what the benefit of religion is, implying that it is indispensable. Many answers can be given:

-it offers clarification when people are confronted with mysterious phenomena, with mysterious experiences and dreams; it explains the origin of everything.

-it offers consolation and is comforting, as it makes mortality less unbearable.

-it integrates the individual in a community and is necessary for sanctioning the social order.

-it offers a foundation for moral values.

These answers are certainly insufficient. Thus, for example, the consolation that is offered is very relative. Though primitive myths explained the natural order and made sense of it, they did not offer any consolation for human mortality. The deceased dwelled in the realm of the shadows where they joined the other ghosts. They were feared by the living, who had to placate them in order to be left alone by them. An eternal soul that would know the bliss of afterlife was totally unknown to primitive people. Even in the great ancient civilisations afterlife, in the sense of a full existence after death, was not evident. Though the Egyptian Pharaoh would live on forever as the Sun God, this was his royal privilege, and it took a long time before civilians who had the means to be embalmed could live on forever. The Hebrews believed that after death they became shadows of themselves and stayed in the Sheol, comparable to the Hades of the Greek. Pythagoras, founder of a new religion, introduced the idea of the reincarnation of an eternal soul that in between re-embodiments went to heaven to stay for a while between the stars. Plato, who was an adept of Pythagoreanism, believed in an eternal soul that was incorporeal, though it was again and again reincarnated as long as its owner did not attain perfection, but when this goal was reached it returned for ever to its birthplace between the stars.

Afterlife as we conceive it was the crucial Christian article of faith, made possible by the sacrifice of Jesus Christ. Originally, however, religious belief did not have the function of reconciling people with death and bringing them consolation. The reason of the universality of religion must lie elsewhere.

Richard Dawkins, looking for the evolutionary benefit of religion proposes the hypothesis that the advantage it gives is not direct. Indeed it is very costly in terms of time and energy that could be spent for survival and reproduction. In consequence religion must, biologically speaking, be considered a waste, which is seldom tolerated in nature. If religion has benefit it must be indirect. He gives an illuminating example. Moths are attracted by light and burn their wings approaching the flame of a candle. This is an indirect consequence of behaviour that is useful for moths. They steer their flight by the light of the sun, the moon, the stars, all objects that are at optical infinity. They are programmed to fly in such a way that the light rays hit their compound eyes at thirty degrees. Doing the same when they are attracted by a candle they fly right into the flame, but the chance that this happens is small, compared to the chance that in their natural surroundings they adjust the course of their flight keeping an eye on celestial objects. Only, we mostly witness the moths that, wrongly, seem suicidal to us and not the hundreds of moths that use star light or moonlight as an accurate compass.[1] In the same way the universality of religion could be the side effect of behaviour that enhances survival. Human infants and young children are taught to obey their parents at once and without further discussion in serious situations. As humans in general and especially young children have a tendency to comply this certainly enhances their chances of survival. They have a natural tendency to believe what their parents tell them in a serious tone.

> Theoretically children might learn from personal experience not to go near a cliff edge, not to eat untried red berries, not to swim in crocodile infested waters. But, to say the least, there will be a selective advantage to child brains that possess the rule of thumb: believe, without question, whatever your grown-ups tell you. Obey your parents; obey the tribal elders, especially when they adopt a solemn, minatory tone.....[2]

The universality of religious beliefs is a vast subject that we shall not further pursue. It must be distinguished from the origin of the similarities in its expression that is of more interest for our subject, namely how men originally conceived of themselves and why it is that in primitive myth all over the world the same themes recur. The latter can be explained by the fact that though each human being is unique, all of them share "the human condition." We are all born, pass through the different stages of life and in the end we all die. Moreover, we have "common human interests," the basic human needs are everywhere similar. Our experience of these basic facts of life is individual, yet it can be supposed that they remain similar in all times, to whichever community we belong and to whichever culture.

A most interesting explanation of the psychology of mythical thought is offered by Joseph Campbell. In the first volume of his vast study of mythology, *The Masks of God, Primitive Mythology*,[3] he tries to make a synthesis of those theories that pertain to his subject. He recognises the major contribution of psychology, especially of the psychoanalytical theories of Freud and Jung, to the understanding of the great myths. However, he will try to give the unconscious, which is presupposed in psychoanalysis, a fresh content derived from scientific data.

This must be seen in the context of the time the work was written. In the sixties of the previous century a new biological discipline developed, namely ethology, the science of animal behaviour. Campbell thought at once it could be useful for his purpose. He was more in particular interested in the study of instinct by N. Tinbergen and in the theories of animal behaviour presented by K. Lorenz. Before that time little had been known about the unconscious level of our mind where our instincts are at work, and now a consistent theory became available. Animals respond instinctively to key stimuli in their environment, and it can hardly be doubted by biologists that humans show instinctive behaviour too. There are, however, important differences. Humans show "neoteny"; they are born before they are fully developed and they retain for some time characteristics of earlier stages of their ontogeny, which makes them "plastic, liable to undergo changes." This characteristic explains that the period during which they can learn is longer than in most other animals. (Neoteny had played a crucial role in the philosophical anthropology of the late nineteenth and early twentieth century. It was used as a key-argument for the ontological uniqueness of our species). Campbell, like many others, stresses that we do not have as many stereotyped, key-tumbler responses as the other vertebrates and that we are less rigidly patterned in our instincts. It is more difficult to chart our instincts than those of other animals, because most of our behaviour even if it is based upon instincts is also culturally determined. Both elements must be taken into account.

Thus Campbell explained the origin of myths in a double way and produced a hybrid account of them. On the one hand he acknowledges the merit of psychoanalytical theories, while on the other hand he attaches great importance to the biological mechanism of "imprinting." Let us start with the ethological part of his explanation. His theory is that our instincts are the basis for the establishment of the imprints images make upon our mind. He opposes this imprinting to the fixed instincts we have, and he calls it "open," because its expression is not totally predetermined. Indeed, the duckling that hatches and shows a lifelong "filial" attachment to the first living being it perceives normally becomes attached in this way to its mother, but if accidentally it is not its mother that is present at the moment of its hatching, but a creature belonging to another species the mechanism works as well. This is exactly what happened during one of the experiments of Konrad Lorenz. While he was watching, a duckling set sight on him when stepping out of its eggshell and from that moment on it followed him everywhere. He concluded that in this way an irrepressible rigid pattern of behaviour based upon an instinct can be triggered off, behaviour that is determined by the impressions of the moment. It is highly improbable that Lorenz, who introduced the term "imprinting," would have agreed with Campbell's extrapolation of the concept that is originally purely biological to cultural contexts. Lorenz defines imprinting as follows:

> a) It is confined to a fixed and very limited period in the organism's life. b) The stimulus situation which will, later on, elicit certain responses, is determined at a time when these responses have not yet matured. c) Unlike all other types of learning, this process is irreversible....[4]

I believe that if we follow Campbell we should understand the term in a looser sense. One thing is certain, what we are taught in our early childhood can provoke emotional reactions during the rest of our lives. Thus in small children who are warned there is a God, who is invisible, but who knows all about us, even our most secret thoughts and all our sins, this idea becomes so deeply engraved in the unconscious part of their psyche, that the mere thought of him triggers off a totally irrational fear. Even for many adults it is extremely difficult to get rid of this reaction.

Imprinting is at work at the unconscious level and its neurological basis is similar in all the individuals of a species, but at the same time it is moulded by individual experiences. This is why Campbell sees in it a plausible basis for explaining mythical themes recurring in cultures all over the world. According to him, in humans imprinting can be caused by events inducing very strong emotions. The peculiar impressions we receive in our early experiences can determine our lifelong immediate response to certain stimuli. Moreover, like many other animals, we do not only react without previous learning to certain stimuli of our environment, but also to artificial stimuli bearing a similarity to the natural ones, and if certain characteristics of them are enhanced, this can make them even more effective. Thus he explains the universality of certain mythological images, acting as "supernormal" sign stimuli, in which he recognises the "archetypes" of C. G. Jung that belong to the "collective unconscious."[5]

He warns us, however, that the fusion of the personal and the universal is different in the psychoanalytical view and in mythology. In the former, he says ironically, it is an endeavour "to translate the whole cultural inheritance of mankind back into nursery rhymes," and this view has not to be taken all too seriously.[6] Mythology is a powerful cultural means that has in the first place the most important function of integrating the individual person into the group and to balance the individual and general interest. Campbell fully appreciates that the use of psychological "complexes," which have their origin in specific traumatic events in the childhood of individuals, as a key for the scientific understanding of the mythology and ceremonies of the group, is "about as appropriate as to mistake a pancake for a souflé."[7] He is not the only one to have criticised, and rightly so, the account of the functioning of our mind and of normal psychological and cultural phenomena in terms of the pathological.[8] He also remains critical of Freud's too exclusive attention for the neurotic in which he translates everything and this into the terms of infantile sexuality; This translation is the predicament of the neurotic, "but if the doctor does so too where do we get?"[9] Indeed, following Campbell Freudians show not only the tendency to put too much stress on sexuality, but also the tendency to rely too heavily upon a regressive explanation, i.e., by referring to the infantile experiences of individuals.

Another point of divergence is the a-historical character of psychoanalytical theories of the motifs of the myths and the rites of different people. These myths and rites are the result of *particular* social experiences and cannot be explained in general psychological terms, because they have in the course of history been adjusted to amongst others geographically determined conditions of existence. They are also, according to Campbell, derived from "millenniums of meditation on the recognized natural order of the living world."[10] Therefore:

> ... in the reading of myth such a [Freudian] reductive method commits us to the monotony of identifying in every symbolic system only the infantile sources of its elements, neglecting as merely secondary the historical problem of their reorganization: pretty much as though an architect, viewing the structures of Rome, Istanbul, Mhenjo-Daro and New York, were to content himself with the observation that all are of brick.[11]

Rather than starting from the pathological, Campbell founds his theory on the most normal experiences. The first experience all human beings share is birth, the traumatic character of which is stressed. The extraction from an existence in the darkness, warmth and comfort of the amniotic water of the womb and the arrival in a totally different world full of light and sounds is a brutal event. Suddenly surrounded by air, there is a brief moment of suffocation, circulatory congestion, then the lungs start to pump and the newly born baby begins to cry. Nobody can recall his own birth, but we can imagine that it must have been a moment of great anxiety. Is there an imprint left in the unconscious layers of our mental activity? This remains a hypothesis, but perhaps it could explain the fact that the birth theme is prominent in the imagery of mythology and religion and that, with a number of variations, it is universally present. It is used not only for the moment of birth itself, but for all moments of abrupt change from one stage of life into the other, from childhood into adult life, from life to death.[12] The rituals that accompany such moments often represent a re-entry into the womb and involve a passage through water. In Christianity baptism, according to *Letters* in the New Testament is explained as a dying, as a going under, as being buried with Christ, followed by rebirth in a new life. It is a participation in Christ's death and resurrection, but also a cleansing from sins.

A quasi universal symbol related to birth, death and rebirth is the labyrinth, often represented by a spiral, which is the image for the womb, dark and silent and where time is lost. To go back into that darkness is frightening: here ends the conscious experience, we are in the realm of death. It can be supposed that in early cultures, youngsters who were about to begin a new phase of their existence and to become full members of the group had to leave their familiar surroundings.

They were plunged in the darkness of caves and left disoriented, lifted out of the context of their former existence. This was certainly a shock therapy, and when thereafter suddenly light was made, the images they saw on the cave walls were forever imprinted upon their minds. They returned afterwards for a while to their mother, but only to be reborn in another, adult life.[13]

The earth as a bearing and nourishing mother is central in the myths of both hunting societies and societies of planters. Nevertheless, the boons she offers must not be taken for granted; she must be treated with respect by observing appropriate rites. The psychology behind this conception is obvious and universal. To the child the mother is at once beatitude and danger. At first it lives in symbiosis with her: she suckles it, gives it protection and satisfies all its needs. Campbell refers to J. Piaget's *The Child's Conception of the World*, where it is shown that mother and child form a continuum, both physical and psychical. According to Piaget the child is at first completely self-centred. There is no distinction between the external and internal poles of its world and all is reduced to what it feels and experiences. Inevitably the moment comes, however, when the mother can no longer anticipate all it needs, and then the child becomes aware that the world does not necessarily respond to its desires. The absence of the mother is felt as threat, loss and collapse of the universe: "Hence, when the mother image begins to assume definition in the gradual dawn of the infantile consciousness, it is already associated not only with a sense of beatitude, but also with fantasies of danger, separation and terrible destruction."[14]

This aspect of the relation of mother and child is reflected in the many myths in which the mother becomes devouring, a cannibal ogress. As soon as the mother becomes a "thou" related to the "I" of the child, her power is thus that though she is life-giving, she can decide to refuse to sustain it or even kill it. In this way the woman, source of all bliss, can become a danger to the male growing into his adult life. She can become a "femme fatale," and the symbol of this cruel aspect of human femininity is the toothed vagina. (Campbell's explanations are generally male-centred. Why the mother should not become a dangerous person in the consciousness of small female children, who certainly are as scared as male children about losing her affection, remains unexplained).

The attraction and antagonism between the male and female element is reflected in other myths. A family structure is biologically required to raise a child, which during the long period of dependence, must be able to rely upon its mother and father. Together with the child, they form a basic unit in human society. The division of labour creates two groups, that of the men and that of the women, and as we can see in primitive communities of hunter-gatherers, the women rear the children, they gather food, roots, nuts and berries in a sack, while the men go hunting and add a small proportion of meat to the diet. The domain of the men and that of the women are separate, and governed by different principles. In mythical thought concepts like birth and death, the male and the female are not only complementary but are opposed. As we shall see in the next chapter, oppositions play a pre-eminent role in myths and religions. To the little boy the centre of his life is the mother, and when he grows up she suddenly changes from being his mother into a woman, and now she can be sexually desired. To the little girl she becomes a rival for the affection of the father, and similarly, the little boy can see his father as a threat, a competitor in the strife for the love of the mother. The tragedies of Sophocles, *Oedipous* and *Electra*, have become paradigms of the expression of these conflicts, which are rooted in prehistoric thought.[15]

Of course there are other universal mythical themes, but from the foregoing can be concluded that in the original way of thinking of men the main human emotions are incorporated, in contrast with modern thought that is governed for the largest part by the emotionally neutral logic of Aristotle. The most important difference is probably that to primitive people there is no clear distinction between subject and object, while Aristotelian logic neatly separates them. Adult Westerners are generally able to distinguish the subjectively experienced and the objectively given, at least in matters of daily life and of science. In the thought of primitive people the borderlines exist too, but they are more equivocal. In their logic an object can be several things at the same time without contradiction. Even grown-ups take it for granted that organisms without a complex nervous system and even inanimate things have not necessarily a soul, but a kind of power, a kind of will. However, this does not mean that primitive people are not capable of gathering accurate knowledge about these same things. However, what they are remains ambiguous, just like in the fantasy of a child during play the broom is no longer a broom, but becomes a horse.

Sometimes the distinction between fantasy and reality is lost and the child takes an ordinary thing for something quite different, like in the following example given by Campbell. A little girl is playing with burnt matches, one of them representing in her game a witch. When she is completely absorbed by her fantasy, she shrieks with terror, because the match has *become* the witch and has the same terrible power.[16] In grown-ups in general, primitive or modern, the borderline between fantasy and reality is seldom, except under special conditions, really lost, and the transition from one domain to another can be made at will.

We can see this from the different roles symbols can play. They seem to be essential for making sense of the world. A true symbol is something that stands for something else, like a flag or an anthem stands for a nation, and can bring tears in the eyes of the true patriot. Nevertheless, nobody believes the flag or the anthem really is the nation. This makes it clear that, though symbols refer in second order to something they do not refer to in first order, they are not identified with it. In some religious or magical contexts, however, the symbol possesses the force of its secondary referent. The representation in a gesture or a picture of an eye is just a representation, but they are also the sign of the "bad eye" and can prevent malevolent persons to cause harm. Holding up a cross or even making only the sign of the cross, and this accompanied by the words "*vade retro Satan*," is merely a gesture accompanied by words, but it has the power to chase the devil away. E.H. Gombrich, in his *History of Art* warns us not to believe that we are completely rid of this ancient mentality. If somebody would give us a photo of a person we are particularly fond of, and ask us that we pierce his or her eyes with a pin, most of us would find it difficult to do so, though we know perfectly well that the picture is only a piece of paper.[17]

We must admit that in the adult the childish mentality is in part conserved. Thus, in a corner of the heart of many of us still resides the idea that reality as a whole is made for us, that we are part of it and that it simply cannot disappoint us in the end. In the words of Campbell:

> We have noted that in the world of the infant the solicitude of the parent conduces to a belief that the world is oriented to the child's own interest and that it is ready to respond to its every thought and desire. This flattering circumstance not only reinforces the primary indistinctness between inside and outside in the child's consciousness, but even adds to this a further habit of command, linked to an experience of immediate effect. The resultant impression of an omnipotence of thought—the power of thought, a desire, a mere nod or shriek, to bring the world in heel—Freud identified as the psychological basis of magic, and the researches of Piaget and his school support this view. The child's world is alert and alive, governed by rules of response and command, not by physical laws: a portentous continuum of consciousness, endowed with purpose and intent, either resistant or responsive to the child itself. And as we know, this infantile notion (or something much like it) of a world governed rather by moral than by phys-

> ical laws, kept under control by a superordinated parental personality instead of by impersonal physical forces, and oriented to the weal and woe of man, is an illusion that dominates men's thought in most parts of the world—or even men's thought in all parts of the world—to the very present. We are dealing here with a spontaneous assumption, antecedent to all teaching, which has given rise to, and now supports, certain religious and magical beliefs, and when reinforced in turn by these, remains as an absolute conviction that no amount of rational thought or empirical science can quite erase.[18]

I quote this text in full, because it explains the psychology not only of mythical, magical and religious beliefs, but also of transcendent metaphysics, i.e., all forms of intellectual activity that implicate wishful thinking and that start from the conviction that there must be more to reality than the reality we experience with our senses. To most human beings the idea that we are born, live and die in a world without common measure with us and completely indifferent to us is unbearable. Therefore it cannot be. There must be something that, if we cannot directly observe it, is nevertheless real and gives sense and meaning to our existence.

An important function of mythical and religious thought is to integrate individuals into society, which in its turn is part of the world order. We must consider this aspect of religious teaching and ceremonial, namely that it is one of its chief aims to suppress as much as possible the sense of ego, and to develop that of participation.[19] Ultimately, the suppression of the self can become, as in mystical experiences, its dissolution, which leads to the unification of the subject with the supreme source of reality, with the One and All, with God. In order to achieve this and to know the bliss of mergence with a superior reality, the believer must first suffer. He must go through an ordeal, he must be cleansed. His spirit must be purified before he can go beyond normal human experience and know the ineffable rapture of beholding "what the eye cannot see." Only then is he worthy to be initiated in the secret knowledge of the men that must not be divulged to the women. The rites of initiation not only mark an important event in the life of the adolescent, they are essential for the integration of the individual in the community. The young child does not participate in the communal life, he is pampered by his mother, plays with other children and as a youngster he shares his time with boys of his age. In more developed societies, the necessary integration is achieved through years of education; in primitive societies it is achieved more briefly and abruptly through rites of initiation. Through these rites, which play a central role in the consolidation of the social structure of his tribe, the youngster is transformed from an individual being, into the incarnation of a mythological, eternal form. The ego of the individual is expanded beyond the biography of the individual:

> A woman gave birth to the boy's temporal body, but the men will bring him now to spiritual birth. They will continue and consummate his post-

> uterine gestation, the long process of his growth to a fully human maturity, refashioning his body and his mind as well, joining him to his eternal portion, beyond time. Furthermore, in the ceremonials that he will presently observe the tasks proper to manhood will in every detail be linked to mythological fantasies of a time-transcending order, so that not only he himself, but his whole world and his whole way of life within it will be joined inseparably, through myths and rites, to the field of the spirit.[20]

Campbell does not consider rites of initiation for women. From more recent anthropological studies we know that such rites exist in various societies. However, not as birth in a new condition, but to make them willing to accept their existence as a woman. There is no change of territory as in male initiations; the rites are performed in or near their domestic surroundings. The initiands do not gain access to secrets guarded by the female community. They are not allocated a totally new role; they have already learned from their mothers how to prepare food, how to care for children and what is the position of married women in the restricted family circle and in society. Once married, they will have to perform the same tasks as before, and this for the rest of their lives. The rites will help them to execute their duties willingly; they are meant to give a cosmic meaning to the monotony of their future existence. It is the total society that carries out the initiation with men and women playing different roles. The men exert a repressive force on the initiand (they may cut her, rape her or force her to work, to run or to stay up all night), the women support her (they may dress her, adorn her, or keep vigil with her).[21] In our society too children and adolescents are educated to become useful members of society, well-adapted citizens, but modern Western culture is profoundly individualistic at the same time. The adolescents are taught that not only society has the duty to provide in the basic needs of its members, but that each of them has a right to achieve his or her personal and private happiness. To care for others, to serve the community, are additional virtues; they add to the value of life, but they are not considered to constitute the ultimate sense of human existence, except for those rare individuals with a vocation to sacrifice their proper interests for an idealistic cause. In other words society is subservient to the individual; the individual has duties towards society, but society is not his or her reason of existence.

The sense of the sanctity of society that is the expression of the world order has existed for an unfathomable long time. It still was present during the Middle Ages, when Kings could be rightfully deposed, if they did not rule according to the eternal moral and social laws ordained by God. From the Renaissance on, the decline of the precedence of society upon the individual begins; the strong bond between the world order and society, and between society and its members, will be broken. Yet, the yearning for a world that is "whole," for a sacred harmony, wherein all things have a natural place and the individual and the general are reconciled, remains, but now the search for a reunion with the sacred and the divine has become an individual quest. Those who cannot believe that for a human being to be born, grow up, raise children, grow old and die is all there is to life, those, who look for something either in this life or in a life after death to give meaning and sense to the fact things are what they are, turn in their despair to religion, to Freemasonry or, if they are more philosophically minded, to metaphysics. As we saw, there simply *must* be "something else," a reality of an utterly different kind in or behind this reality. On the other hand, Western science gradually developed, and from the nineteenth century on, when education was generalised, an important proportion of the population was familiarised with at least a simple version of the well-established scientific theories of nature, of life, of man. Science, however, offers no comfort for those who have an intellectual, emotional and moral need to feel integrated in the totality of being. Some believe, wrongly in my opinion, that science can be reconciled with metaphysics and religion.[22] The methodology of science forbids that metaphysical or religious conclusions are drawn by extrapolation from its theories *by scientists in their role of scientists.*

In the twentieth century society is secularised. The sacred can be found in religion, in mystical experiences or in metaphysical speculations, but in any case it is not nearly as vividly present in daily life as it was in the past. In countries where it is guaranteed in the constitution that church and state are separate, belief and faith are considered to be purely personal matters, determined by family tradition and upbringing. In this, as in other respects, our society is atomistic. Though rites of initiation are still practised in a religious context, they no longer serve the purpose of integrating a person in society as a whole, but only the integration in the religious community to which he or she belongs. Thus in the Catholic Church at the age of six is celebrated the "first communion" of the young boys and girls and at the age of twelve follows the "solemn communion." The equivalent for the Protestants is the "confirmation." The Jews celebrate the bar mitzvah, when their boys are twelve, and the Muslim community observes a similar rite. These rituals mark the entry of the adolescent into the religious community, but they are a bleak reflection of the much more sweeping rites of initiation of primitive societies.

Primitive religions are "monophysisms," which means there is only one reality, comprising natural phenomena and supernatural phenomena. The rites of initiation are meant to realise the incorporation of the adolescents in the restricted community, where everybody knows everybody, and at the same time they are meant to incorporate them in the world order. In other words, these initiations have a cosmic and universal dimension. To primitive people the world is their world, the world they know. It has no complement in a transcendent reality. It is explained by a sacred myth, and the initiation not only transforms individual youngsters, who were hitherto spoiled brats, into genuine members of their community, but sometimes they are reborn into incarnations of mythical personages with a timeless dimension. As we have seen, the experience of the initiation is frightening, often very painful and of such a nature that the word transformation can be taken literally. The boy who undergoes the rite suddenly becomes a man sharing the secrets of the adult men of his tribe. He is now introduced in the realm of the sacred.[23]

Not in all primitive societies is individuality experienced in the same way. J. Campbell stresses that there is an important difference in approach between societies of planters and communities of hunter-gatherers. A possible explanation is that planters are much more dependent upon the Gods of the elements, because droughts or periods of too much rain can cause the loss of the crop and result in famine. Therefore, obtaining the benevolence of the Gods is of vital importance and the life of the community is centred upon the rites and ceremonials serving this aim. Most of the time of these peasants is taken up by their labour in the fields and by their religious duties. The latter require much of their attention, the religious calendar must strictly be observed, the exact words of the rituals memorised, all the details of the procedure executed with extreme care. This means that the larger part of the adult life of these farmers is spent in labour and in common activities and not much time is left for individual creativity.[24] It is a well-known fact that rural communities are conservative and seldom at the vanguard of progress. Anthropologists are not far from assessing the existence of hunter-gatherers, as compared to that of agricultural people, as paradisiac, chiefly because they have much time for leisurely activities, for socialising and feasting. Nature provides in the basic needs of the group: hunters are far less dependent upon the weather, and hunter's luck is very different from the whims of the Weather Gods. The hunter is a personality with distinctive qualities, not the incarnation of a perennial mythical personage, and he functions far more in the group according to his personal dispositions. Campbell, in order to illustrate this, contrasts the priest of planter societies with the shaman of the hunter-gatherers:

> The priest is the socially initiated, ceremonially inducted member of a recognised religious organisation, where he holds a certain rank and functions as the tenant of an office that was held by others before him, while the shaman is one who, as a consequence of a personal psychological crisis, has gained a certain power of his own. . . .[25]

The spirits that visit the future shaman during the period in which he withdraws himself from the others, fasts, and submits his body to extreme cold or heat or other agonies, are his personal ghosts, and what he is going to discover is unknown beforehand. The majority in the group, though, are "honest hunters"; each individual has his own character and skills, revealed to him in puberty rites based on their myths, which uncovered his nature, but these people are not to be compared to the shamans. The shamans are individuals singled out by a special gift to reach in a state of trance a layer of reality forever hidden to the ordinary members of the tribe. They have special powers, can heal people, can harm people, they have visions. They are respected, but also feared by the community, which can turn against them. Indeed, the belief they have special powers leads to holding them responsible for the evil that befalls individuals or the community, which can result into taking revenge on them.[26]

The contrast with the culture of the planters is obvious. The myths, rites and ceremonials of the rural society are designed to suppress manifestations of individualism; the individual is taught to see himself as a seed to be sown, buried in the earth in order to give birth to new life. Acquiring the benevolence of the Gods, enticing them to contribute to the abundance of the crop and a rich harvest, is a condition for the survival of the community. Hunter societies, on the contrary, have advantage to develop in their members individual qualities, by encouraging them to discover their true personality, to be capable of self-realisation. No corn-Gods, no cloud-Gods rule over their existence, they are not a thread in the pre-established pattern of a social and cosmic tissue, but these individuals are courageous men, who can bear suffering and solitude in which through supernormal experiences they discover their true self. The religious life and myths of the hunters are less developed than those of the planters, yet, they cherish their sacred traditions. It is the shaman who is their guardian and the reciter of their chants.[27]

These reflections on the psychology of primitive thought reveal the sources of religion and metaphysics that give meaning and sense to reality. In primitive thought religious and practical knowledge are inextricably interwoven, and in our culture they became only recently separated. Campbell contrasts Job and Prometheus: the former, like the peasants of the primitive societies did for the Gods of the elements, bows his head for the Creator, even though God has treated him unjustly and wronged him, the latter revolts against the Gods, challenges Zeus, even deceives him. Prometheus is the strong individual who steals fire from the heavens, and gives science to the mortals, who lived hitherto in ignorance. He is a super-shaman; he is related to "the great trickster," a mythic personage, who represents chaos and disorder, but who is also the giver of great boons and the teacher of mankind. Prometheus is cruelly punished for his rebellion by Zeus. He is chained to a rock high in the mountains of the Caucasus; in daytime his liver is eaten by an eagle, in the night it grows again, so that his torture goes on and on. In spite of his suffering he is certain that one day his agony will end, the day when his chains will fall off of themselves and the reign of Zeus will come to an end. Prometheus possesses titanic power and this makes him fearless of God; he knows the laws of nature and this same knowledge will one day enable man to dissolve the shackles of religion.[28] The contrast is strong: on the one hand the manly force of the strong individual, on the other submission and womanly piety.

The difference between the two types of culture that is highlighted here is probably exaggerated. Nevertheless, it is worthwhile to reflect upon it as a possible source of the remarkable duality in Western culture of which two components, science and religion, which in other cultures merge with one another, are here conflicting elements. It is well-known that at the end of the Middle Ages the cosy picture we had of the world, of our planetary system and the space surrounding it, changed dramatically. Step by step it was modified each time a new theory of our planetary system was proposed. Cusanus, Copernicus, Galilei, Kepler, Newton contributed to the destruction of our old certainties. The role of God became far less important, until he became a superfluous hypothesis for science. However, biology did not progress as rapidly as physics and astronomy, and in consequence our conception of ourselves did not undergo an equally drastic change. It remained extremely difficult to explain in a naturalistic way how life came about, let alone how we came about. Life had to be created or at least its blueprint had to be of divine origin. Even if we were not directly created by him, for centuries we still imagined ourselves to be the seal on God's work. It is only when Darwinism became generally accepted that the scientific point of view and religious as well as metaphysical theories of man came to a final clash. However, the urge to believe that there is a transcendent foundation for our existence, giving it meaning and sense, has remained in the hearts of the vast majority of men, even if science cannot confirm this belief. In science wishful thinking has no place, and therefore it often goes against the deepest cravings of men, against that part of our psychology that is the adult version of the infantile confidence that in the end all will be conform to our needs and desires. Once this confidence lost, it is difficult to believe we were preordained as a necessary part of a harmonious world order or that we have the mission to complete reality conform to a divine providence. If from a scientific point of view our species is the product of a mixture of chance and necessity, like Jacques Monod proclaimed, it is no longer part of a pre-existing plan. Our existence has no longer a cosmic significance.

Notes

1. R. Dawkins, *The God Delusion*, Black Swan, London, 2007, 201–202.
2. *Ibidem*, 203.
3. J. Campbell, *Masks of God*, I, *Primitive Mythology*, Penguin, Arcana, Harmondsworth, 1991, (reprint of 1969 ed.)
4. K. Lorenz, "The Role of Gestalt Perception in Animal and Human Behaviour," in: *Aspects of Form*, Lancelot Law Whyte, (ed.), Lund Humphries, 1968, 157–178.
5. Campbell, *Masks of God*, I, 32.
6. *Ibidem*, 64.
7. *Ibidem*, 92.
8. D. Morris, *The Naked Ape*, Jonathan Cape, London, 1967, 10–11.
9. Campbell, *Masks of God*, I, 65.
10. *Ibidem*, 91.
11. *Ibidem*, 65.
12. *Ibidem*, 61–62.

13. *Ibidem*, 66–67.
14. *Ibidem*, 68.
15. *Ibidem*, 76–77.
16. *Ibidem*, 22.
17. E.H. Gombrich, *The History of Art*, Phaidon, 1986, 20.
18. Campbell, *Masks of God*, I, 80–81.
19. *Ibidem*, 82.
20. *Ibidem*, 89.
21. F. Bowie, *The Anthropology of Religion*, Blackwell, Oxford, 2002, 172–176.
22. L. Apostel, "Can Metaphysics Be a Science," *Studia Philosophica Gandensia*, Ghent, 1963, 7–95
23. Campbell, *The Masks of God*, I, 89.
24. *Ibidem*, 229–231
25. *Ibidem*, 231.
26. *Ibidem*, 249.
27. *Ibidem*, 250.
28. *Ibidem*, 279–280.

Chapter 2. The human individual in primitive thought

Introduction

The logic of primitive thought

As we have seen, primitive people are "monophysists," which means that reality for them is not composed layers of different kinds of being, but that nature is one. The primitive world view is not anthropocentric; to be human is to be one of the parts of reality on a par with the others and to relate with them, rather than to be ontologically different from all that exists. The world comprises the sun, the moon, the stars, rocks and rivers, plants, beasts and men, Cultural anthropology teaches us in the first place that our modern view upon man is by no means the only possible one. We do not know how peoples that vanished from the earth in the far past thought about themselves, but we know for certain that in "primitive" cultures that have been studied by anthropologists in the recent past, the way men perceived themselves was totally different from the self-conception most Westerners of today cherish. "Savage" thought, in spite of the fact that remnants of it can be found in our ways of reasoning in certain contexts, is in many respects at the opposite of our present-day mentality. It is not only the content of these ways of thinking that is different, but their respective logic, which is the reason that they are incompatible with each other. We seldom are aware of our own conceptual scheme, of its rules of classification and valid reasoning, i.e., taken in a broad sense, its logic. We have internalised our ways of thinking when we were small and taken them for granted, but they are by no means eternal and universal, as becomes clear from anthropological studies. In fact, our modern approach to reality is only a very recent development preceded by a much longer period in which first the primitive and then the great ancient cultures developed and thrived. To complicate things further the domains of our cognitive activities are differentiated. Our psychology allows us to switch from one type of thought, in use in one domain, to other types, valid in other domains, without much effort. From the way we think in daily life and treat practical matters, we can pass smoothly to theoretical, scientific thought, and apparently it costs many of us very little effort to enter the domain of religious contemplation or metaphysical speculation, in which remnants of the ancient mentality are still present. Curiously enough most people are not conscious of the fact there is an incompatibility in these different ways of thinking. Both religion and metaphysics, however, have their roots in the primitive, mythical mode of thought, and it is not at all obvious that their combination with common sense and scientific knowledge does not lead to serious contradictions.

Like all other living beings of one species, men resemble each other in a great number of respects, but each person is nevertheless unique, because each characteristic is realised in him or her in a singular way. Human beings have been aware of this from times immemorial and they have tried to understand themselves both as individuals and as sharing a common "nature," with however many variations they defined the latter. Moreover, humans are social animals; they live in groups, tribes, societies, never alone, except by some rare accident, like the shipwreck of Robinson Crusoe. We are programmed by our various instincts for social life, which means that we are social from the moment we are born and do not have to learn it. Nevertheless we do not live in a pack or a flock with a quasi invariable structure, but in societies of which the many different forms are to a large degree culturally determined and do not correspond to an invariable order based on instinct alone. From the beginning our forebears were not only aware of the contrast between nature and culture. They had the intellectual need to make structured representations of both realms that paralleled each other, were coordinated and kept each other in balance. Because the social structure reflected the world order, it was sanctioned, eternal and necessary.

Their world view was in part based on empirical knowledge and at the same time it was religious; it gave meaning to all aspects of their life. Because of its sacred character it was transmitted with very few alterations from one generation to the next one. Anthropologists have tried to reconstruct the culture of the hominids, our remotest ancestors who are lost in time, from what they came to know about tribes still living in the Stone Age. It is doubtful that we can gain any certainty about how the hominids, even how early *Homo sapiens* really thought; also the interpretation of the archaeological remnants of the different hominid cultures is a delicate matter. It would, therefore, be inappropriate to understand the term "primitive" we use here as referring to the earliest hominids. Lévi-Strauss saw even in the material he gathered when staying with Amerindians (1939–1941) only the remnant of the far richer cultural heritage that had existed in the past.

It is with this restriction in mind that we must consider what follows. Claude Lévi-Strauss, the father of anthropological structuralism, has tried in *La Pensée Sauvage* to lay bare the structure of the original forms of human thought as it could be discovered in the culture of so Perhaps a biological and psychological explanation of the genealogy of the almost universal images and patterns of primitive thought is possible after all, but not in terms of inborn structures. In the previous section we have seen that Joseph Campbell has tried to account for the many common characteristics in primitive mythical thought all over the world, by relating them to biological and psychological data that are at once universal and leave room for individual experience—called "savage" or "primitive" men of his day. When he wrote this book, the idea that "savage" meant that they were not only crude, but also simple, had been dismissed for quite a while already. During his investigations Lévi-Strauss saw confirmed the more enlightened thesis that, on the contrary, they were capable of constructing sophisticated conceptual systems with a rigorous logical structure and requiring the capacity to think in highly abstract ways.

As the ontological status of structures can be a matter of contention for the scientist, it is important to note that Lévi-Strauss rejected the Platonist interpretation and replaced it with a more Aristotelian notion of immanent order. He considered philosophical speculation as a sin in purely scientific work. Influenced by F. De Saussure, he applied the method the latter used in his structural linguistics to cultural anthropology in order to give it a scientific backbone.

As he explains, i.a. in *Tristes Tropiques*, the most adventurous part of his work, the exploration full of dangers and discomforts of hitherto undiscovered territories, where tribes of Indians lived who were never visited by white men, an activity that laymen imagine to be the essence of the trade of the anthropologist, is in fact a necessary evil. The gathering of empirical material is not interesting *in se*, but can at most result, once the anthropologist is home again, in series of popular lectures illustrated with photos and films. The serious work is done in the study of the scientist, at his worktable, where he can sort out the material and compare the intellectual products of the different tribes. Then, digging like an archaeologist, he must try to discover under the variable cultural "superstructure" the basic universal mental patterns or the "infrastructure" that gave rise to it. As he attested, next to De Saussure, Marx and Freud were his main sources of inspiration. They taught him to forget about the seemingly obvious and to look for *hidden causes*. The results obtained by the use of this "archaeological" method have been most impressive and they have great explanatory value.

Reflecting on the ways humans conceptualise the world and conceive of themselves we want to explain them. Though probably in early stages of cultural development a primitive type of logic must have been used everywhere, the structure of conceptual schemes and their logic is variable and in consequence they cannot be inborn. The existence of basic linguistic structures, universal characteristics of all languages and a primitive kind of logic can no doubt be attested, but it was never shown that they are due to a particular form of organisation of our brain. As we shall see, the switch in our culture from the ancient logic to the new logic of Aristotle has caused a dramatic change in our mentality that eventually made possible the development of modern science. Nevertheless no corresponding important modification has taken place in the way our brains are constructed.

Perhaps a biological and psychological explanation of the genealogy of the almost universal images and patterns of primitive thought is possible after all, but not in terms of inborn structures. In the previous section we have seen that Joseph Campbell has tried to account for the many common characteristics in primitive mythical thought all over the world, by relating them to biological and psychological data that are at once universal and leave room for individual experience.

In my personal opinion our thought is layered, each layer reflecting a phase in our intellectual development. All humans have evolved from common ancestors; we inherited the same instincts, drives, emotions and inborn mechanisms, and we once all shared relatively similar living conditions. Therefore it is probable that we perceived and experienced our environment in similar ways and that we also shared, and still share, what is called "basic common human interests." The fundamental stage of our cognition, our initial "experimental" understanding of reality is built during the first years of our lives, i.e., during a prelinguistic phase, and is soon followed by the next stage in which we learn a language. Meanwhile, and to a large degree independently of the development of our linguistic skills, our minds mature and we become ready to absorb step by step the conception of the world order and of the social order shared by the members of the group of which we must become over time full members. This will constitute the third layer of our intellectual equipment.

We have still no answer to the question why on the past we all started with the same type of conceptualisation of reality. Though there is no real proof for this, it seems to be at least a sound hypothesis that the fundamental similarity of human experiences and emotions are sufficient to explain the recurrence of a great number of mythical themes and their patterns in so many cultures, and that no need exists to introduce unwarranted, purely theoretical entities.

About the logic in primitive thought

As we have seen, primitive people are "monophysists," which means that reality for them is not composed of different kinds of being, but that nature is one. The primitive world view is not anthropocentric; to be human is to be one of the parts of reality on a par with the others and to relate with them, rather than to be ontologically different from all that exists. The world comprises the sun, the moon, rocks, rivers, woods, living beings, but also spirits and Gods. Being a component of this totality, men wanted to know it in its details, but also to understand it as an eternal general order, an equilibrium of antagonistic principles. Indeed, adequate knowledge is of course necessary for being able to survive in it, but it is just as important to make sense of it and by reconstructing it mentally as an ordered totality to feel at home in it.

The lowest and most fundamental level of our knowledge is based upon our faculty of perception, and it enables us to understand and catalogue what has been observed. Next to this, there is the need to grasp not only *how*, but also *why* things are as they are. It is natural for these primitive men to regard the relations they have with the various elements of their environment as no different from the relations they have with other humans, and thus they make sense of natural phenomena by interpreting the unknown by the known. This is not to say they do not distinguish nature and culture, but for them the elements of nature are "thou," i.e. subjects they are confronted with, endowed with emotions and capable of acting and reacting like we do, and having concerns very alike to our own.

Moreover, whatever can exert power is real. Thus a religious object representing a sacred animal, a spirit or a god, can not only function as a symbol, but it can take quite literally their place. Thereby the object becomes active; it is endowed with a force that is very real and can be felt. This does not mean all things possess a soul or "anima," but it means that for these people there is no sharp distinction between subject and object, between what is animated and what is not, what has an emotional value, negative or positive, good or bad. We distinguish sharply between efficient causes, which to us are most important, and final causes, i.e., the intentions of living beings, which are only in a secondary sense to be considered as causes. Though it was once believed that primitive men had no real conception of causality, this is wrong; it is only a different one from ours, being more equivocal.

Central myth

In primitive thought the objective and subjective approaches can be combined, and as a consequence, how things appear depends upon the chosen point of view. Empirical knowledge and religion are not yet separated. The knowledge of these men, encompassing both, consists at the same time of recording what is presented to the senses and understanding the meaning of what they observe by referring to a "central myth" and to myths and legends related to it. This central myth gives an account of how the world came about; it explains why natural phenomena are as they are, and also how they are related. Their mythical cosmology establishes an exemplary order that is projected upon the actual, concretely experienced reality with its overwhelming variety of elements, and turns it into a continuous comprehensible system.

For a long time it was supposed that primitive people were not capable of abstract thought, and that they were only interested in the use they could make of what they found in their environment. Lévi-Strauss has shown that this is a mistake; though they depend for their daily needs upon nature and are certainly interested in its purely material aspects, they have also intellectual needs that go far beyond this. That they are capable of sophisticated forms of abstraction can be seen from the fact they create conceptual systems combining elements of a wide variety of cognitive domains. In order to illustrate this we can consider the conceptual scheme behind the seasonal rites of the Pawnee Indians. The poles erected at the corners of the hut where the celebrations take place are made, according to their orientation towards the four winds, of different kinds of wood and they are painted in contrasting colours.

The scheme below (Table. 2.1) shows clearly that primitive men relate *concepts from different domains* of thought and switch without transition from one of them to another one. It comes as a surprise how smoothly they can pass in this way from living beings, like plants, to colours, from spatial concepts like wind-directions, to temporal concepts like seasons. The scheme also shows the fundamental role of *associations based on pairs of opposites*: north-south, east-west, summer-winter, time-space. Their peculiar approach to reality relating all the elements to each other does not only allow them to link concepts that are disparate by their content, but also to pass without transition from the least general concept to more and more general ones; they switch from species to groups of species, then to a system of properties or categories. In this way their universe is a continuum of successive pairs of contraries.

SPACE	poplar......white.....*south-west*	south....*summer*	TIME
	negundo.....red......*south-east*		
	elm.........black....*north-east*	north.....*winter*	
	willow....yellow....*north-west*		

Table 2.1 Cf. Cl. Lévi-Strauss, *La Pensée Sauvage*, Plon, Paris, 1962, 185.

The general "elasticity" of the conceptual framework of primitive men is astonishing. We encounter in their thought local systems of classification in the overall system. These local systems can be broadened and cover domains that are outside their initial reach and that are very different from it, and sometimes they can even reach a universal dimension, or on the contrary, local classifications can shrink to the dimension of concrete things or individuals by a conceptual process of particularisation. Lévi-Strauss gives an enlightening illustration of this phenomenon of plasticity found in the culture of the Aranda, aboriginals living in a rocky region of Australia. These people recognise in the peculiarities of their land the concrete realisations of elements mentioned in their myths. At the same time they identify each of these concrete landscape features with a mythical reality that corresponds to different phases of their rituals, mythical reality that thereby determines the programme of their ceremonies.[1] We can see from this that the succession of the ritual phases in their social time and the general mythical geography of space correspond to each other. As we remarked earlier, the transition from concepts related to time to concepts related to space seems to be natural to them, but other domains that are unrelated to our mind are also closely associated to theirs.

The way primitive people conceive of themselves cannot be understood without a brief analysis of the rules they use for classification. Let us enumerate a number of characteristics of their peculiar ways of classifying things. (It must be remarked that they apply the rules without being conscious of them and without being able to formulate them explicitly, just like we are able to speak in a grammatically correct way without being necessarily capable of indicating the rules. Both the canons for classification and for constructing a conceptual scheme and the canons for correct grammar have been developed over an indefinite time by what must have been countless generations. Let us consider the following example of classification from the Hanunoo Indians.

Entity	=	**something that can be named**
Thing	⇔	neither a person nor an animal
Plant	⇔	not a stone, etc.
Herb	⇔	not a ligneous plant
Pepper (*Capsicum sp.*)	⇔	not a rice plant, etc.
Domestic Pepper (*Capsicum annuum L., .., etc.*)	⇔	not a wild pepper
Chili Pepper	⇔	not a green domestic pepper
Cats Penis	⇔	not one of the other terminal taxa

Table. 2.2. Cf. *Ibidem*, 184.

We can derive a first characteristic from the classification scheme in Table 2.2., to wit that there it is *distinctive*: a thing is characterised by *not being* something else. By this classification we do not find what is similar, e.g. all pepper plants, on the same level, but nevertheless the classification is sound according to our norms.

A second characteristic is that an order based upon natural elements can be constructed and then *projected* like a grid upon an aspect of human reality. For example, each clan of a tribe of the Osage possesses a sign of life, a totem or divinity, that lends it its name: puma, black bear, young dear, royal eagle, etc. (In fact this totem can be anything that remains more or less constant in time and can stem from the conceptual, moral or physical domain). All of these signs belonging to clans are supposed to share a common property, e.g. a colour that is found on one of its parts: it can be the black colour on the underside of its paws for one kind of animal, or the tip of its nose for another one, or on the tips of its ears, or on the end of its tail, etc. Each animal is thus decomposed into its parts following a law of correspondence (snout = beak, etc.) and the equivalent parts are regrouped by the common characteristic, namely in our example the presence of a part that has the colour of coal. (The property reflects their belief in the sacred power of fire on which certain rites are based. For example the warriors must blacken their face before going to war. If they omit to do so, they have no right to recount their glorious deeds and to receive military honour).[2] This peculiar conceptual approach shown in Fig. 2.3. permits *to pass from one logical level to another one*, from the categories to the species showing the property in question, and from the similar elements in individuals to the species. Each species is mentally divided into its parts, and this division permits to re-establish the totality on another level. In fact a preconceived "grid" is applied to any empirical situation showing enough affinity in order that the elements that are obtained preserve certain general properties.[3]

In consequence in this logic, *the relation between what is individual and what is general* is completely different from what we are accustomed to. The following simplified diagram of such a type of system will make this even clearer. The explanation of the diagram in Fig.2.3 is the following. First come empirically based entities, namely species (reduced for the sake of simplicity to three): species Seal, species Bear, species Eagle. Each of them contains a series of individuals (equally reduced to three for each species), seals, bears, eagles. Each animal can be analysed into its parts, all the heads, the necks, etc. A last regrouping results in the model of the reconstituted individual. This is the general scheme that can be found in totemic classification. However, in reality these schemes are not only much more complicated, but present a wide variety. Also, Cl. Lévi-Strauss writes that a classification of all these variants of complications is conceivable, but would be so complicated that this would require machines to do the necessary computations.[4] (This was written in 1955).

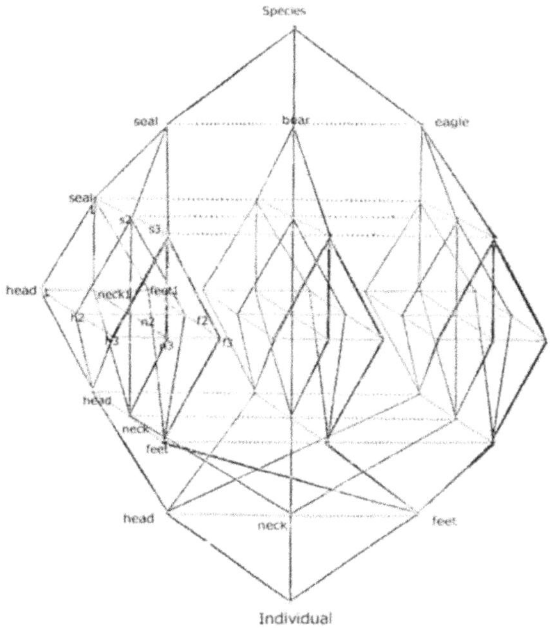

Fig. 2.1. *Ibidem*, 200.

Self-conception

Now that we have an idea of the conceptual schemes used in "savage" thought, we are able to see what the implications are for the way primitive people conceive of themselves. Let us consider the case of the Bororo.

The organisation of the society of the Bororo Indians from Brazil is a most enlightening example for understanding how primitive people create their idea of what it is to be a person. The Bororo in the past all built their villages in the same way that mirrors the order in their surroundings and has a sacred meaning. This becomes clear from Fig. 2.4. The huts are disposed in a circle with in the centre the large house of the men. At the eastern side of the central house is the oval dance floor. The circle is divided into two halves following the north-south direction. The halves are reserved respectively for the group of the "Cera" and the group of the "Tugaré." The nearby river is tangent to the circle and the village is divided a second time, by another imaginary line from east to west. (The concrete realisation of the plan is only an approximation of its abstract structure). Thus the inhabitants of the village are again divided into the upstream people and the downstream people. The meaning of this second division is not totally clear to Claude Lévi-Strauss, but probably it has something to do with the dwelling of the dead in two separate realms. The construction thus resembles a cart-wheel, with as the knob the house of the men, and as the spokes the paths leading from the house of the men to the huts marking the circumference. This illustration shows a vital feature of the organisation of the social life of these Amerindians, organisation that is based on sacred rules and cannot be disrupted without destroying the social tissue. The huts belong to the women, who inherit them from their mothers. The men are Cera or Tugaré by their mother, and they marry always a woman from the opposite group; thus a Cera man only marries a Tugaré woman and a Tugaré man a Cera woman.

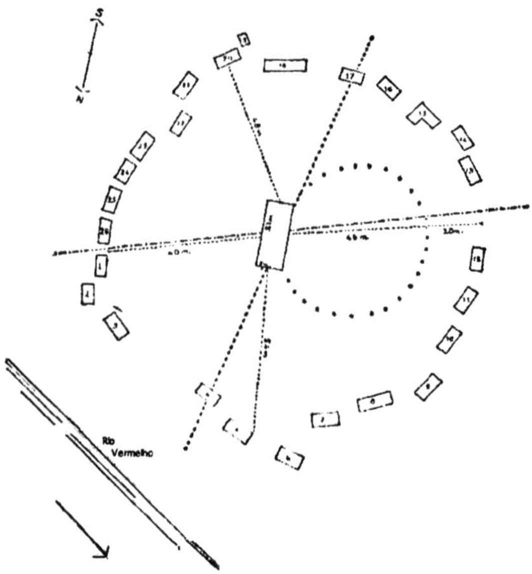

Fig. 2.2. Lévi-Strauss, *Tristes Tropiques*, Plon, Paris, 1955, 188.

When getting married, a man leaves the house of his mother, where he was brought up, goes over to the other side of the village and now lives in the house of his wife. He remains of course welcome in his former home, but if it seems convenient for some reason, he can go to the house of the men, which is both Cera and Tugaré because situated on a terrain overlapping the two halves. The house of men is forbidden for the women, but there a man can freely chat with his comrades, sleep and eat. This system neutralises the opposition between the two halves. Paradoxically the Cera possess the political and religious power, but are called "weak," the Tugaré have power over nature and are called "strong." The reason is that the Cera rule over human affairs, while the Tugaré according to the central myth of the Bororo were in the past creators and Demiurges ruling over nature.

From the Tugaré group are recruited the "bari" or magicians responsible for the pathways followed by the stars, and responsible for the wind, the rain, health or illness, and finally death. Another way of equilibrating the opposing halves is that all the duties of a person profit the other half of the community or are accomplished with the help of its members. Thus the burial of a Tugaré is conducted by a Cera, and vice versa. In fact, the two halves of the village are partners, and in the words of Lévi-Straus each social or religious act of a person implies the assistance of the vis-à-vis who plays the complementary role of the role that is assigned to him.'

There are further divisions: the population of each half is partitioned into clans, each clan is divided into a number of families. In addition there are in each group of clans red and black clans. Another distinction cuts through the ones previously mentioned: all the members of the population belong to the high, the medium or the low class. This leads Lévi-Strauss to draw the conclusion, that the marvellous conceptual structure, which is projected upon the social life and regulates it, in fact is used to sanction the more crude reality of daily experience. Indeed, the "good savage" is, as could be expected, an idealisation, just like a purely egalitarian society of men living in a kind of paradise. In fact, there is an important difference in wealth between the social classes, which becomes most evident from the contrast between the rusticity of the possessions of the lowest class and the refinement of those of the upper class. Fig. 2.3. illustrates the complexity of the social organisation

Now we are able to answer the question "what does it mean for these people to be a person?" It is in the first place to be a Bororo. In the second place it means to be the member of a community living in a certain village. In the third place it means to belong to the Cera half or to the Tugaré half, and in consequence to marry a person of the opposite half. In the fourth place it means to be a member of the group of the upstream people or of the downstream people. Further, and in the fifth place, it is to belong to a red or black clan of one of the halves and in the sixth place it is to belong to one of the families of which the clan is composed. Finally, it is also to belong to the upper, middle or lower class, as shown in Fig. 2.5. It is clear from this that the individual fills in a unique place in the overall social structure, and knows exactly the characteristics of his or her place by which he or she is defined. Of course, but this is not explicitly mentioned by Lévi-Strauss, to be a man or a woman makes a world of difference in terms of lifestyle and privileges. A most significant difference is that the women are not initiated in the religious mysteries. (The comment of Lévi-Strauss is that this compensates for the power the women have by owning the huts they inherit from their mothers.)

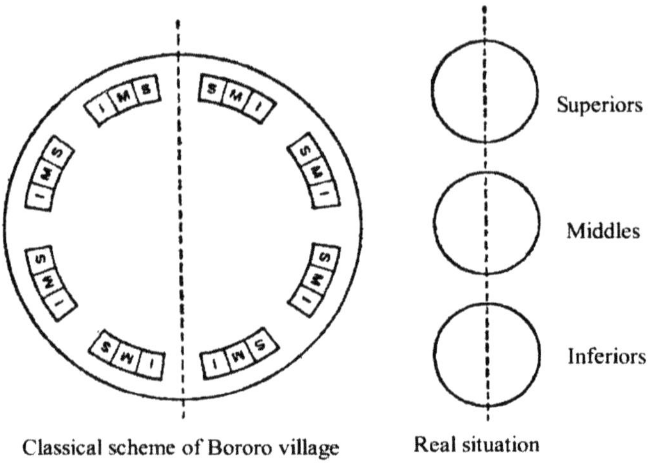

Fig. 2.3. *Ibidem*, 211.

Each person can be compared to a piece in a chess board game: it is either white or black, a king, a queen, a knight or a bishop . . . ; each of these chessmen is a constitutive element in the total game obeying a number of specific rules. For a Westerner of the 21st century it is difficult to imagine the full implications of such a kind of social life. In our society, though a person is of course not born in a social vacuum, but in a definite social situation, in general an individual is expected to develop itself, profiting from its qualities, overcoming its faults, grasping the opportunities that are offered, etc. He or she is encouraged "to become someone," which means that he or she is not already predestined to be this or that, but must make choices, and must make something out of his or her life. The general belief is that "where there is a will, there is a way." The responsibility of a person in a primitive society is to correspond to his or her position in the system, while people of our time are thought to be morally responsible for what they are, i.e., for what they have chosen to make of themselves.

W. K.C. Guthrie observes that in early societies, which were small, there was no conflict between moral duty and self-interest; compliance when one is supposed to follow the rules brings personal satisfaction. He cites Ritter who remarked in this context:

> He who in his relationship to his fellow men and the gods observes the existing customs is praised, respected and considered good; whereas he who breaks them is despised, disciplined and considered bad. In these conditions obedience to law brings gain to the individual, whereas transgression brings him harm. The individual who obeys customs and law is happy and contented.[5]

Many biologists have warned us that once we were hunter-gatherers living in small communities where everybody knew each other and where everybody knew his or her own place and rank. The members of the group had no doubt about what could be expected and demanded from each of them and how they had to be treated. A profound sense of belonging to a family, to a clan, to a tribe or group is natural for humans, probably because we once hunted in packs with a tight hierarchical structure. In contemporary society, it is often stressed, more and more people live in more and more densely populated towns and they have become, paradoxically, extremely lonely. Our society is certainly highly organised, but we do no longer understand ourselves from its totality and its structure. We are mainly related to it by our work, which assures a living for ourselves and for our family. In our welfare society our basic needs are secured by the state, and in our democratic systems, we can once in a while express our opinion by our vote, yet, we do not understand who we are by referring to the global population, which has become to a large degree an abstraction.

In our Western societies we do not come into the world with a ready-made identity, we must construct it. In large cities most people, outside a small circle of friends and acquaintances, are anonymous. Families are far less stable than in the past, i.a, now that divorce has become so common. The role of man and woman in a household are no longer stereotypes. It is much more difficult for adolescents to think of their parents as paradigms, and in consequence, youngsters frequently suffer from an identity crisis. The symptoms of this estrangement are recognised by sociologists: the success of youth gangs, of various sects, of football fan clubs, signal the lack of integration and of a sense of security. All this is well-known. In the distant past many individuals when the burden of life became too heavy sought refuge in cloisters, where they led a life in a religious community with strict rules, ceremonies and rites, abandoning their personal ambitions. They lost their autonomy and they had to serve the community, but on the other hand they were freed of the heavy burden of being solely responsible for themselves. In addition, their existence, however austere it was, at least made sense. It is not without significance that when entering the order, the brothers chose a new name, mostly that of their patron saint, giving up by this fact their old identity.

Name-giving

Name-giving in primitive society can give us further evidence of the distance between our own mentality and that of primitive people. Name-giving is *a special case of ordering reality by classification*, which Lévi-Strauss considers to be the great cultural feat of Neolithic man. Totemism and totemic name-giving is in its turn just a peculiar brand of this general phenomenon. We have seen previously that general and particular are *relative* in primitive thought, as becomes clear from some characteristics of the names the natives use for human beings in general and for persons. In most cases only members of the own tribe are considered to be humans, members of other tribes are "barbarians," less than humans or even devils. In contrast with this chauvinism, in other cases we can find a tendency towards the broadening of the concept of humanity and this to the extreme. Of the latter tendency examples can be found in the "international" classifications of clans, totems and divisions of totems. What is more, in some cases even domestic animals can be included in the classification of beings that are considered as humans and this by giving them totemic names.[6] In addition, the conceptual scheme of Neolithic man can undergo *transformations*: when used for generalisation, the mazes of the structure can be widened and the network stretched to a degree where it becomes very loose; when on the contrary used for particularisation, the mazes can be tightened and become very small covering only a very specific domain.[7] In the matter of naming we see the same flexibility: general names can be used to individuate, proper names can be used to classify somebody under a broad category. (We can recognise the same phenomenon in the way we still use names for persons. Let us suppose that two different persons, somebody's brother in law and his best friend, have as a Christian name John, then often the family name is used to specify which John is meant, but conversely the Christian name is used to specify which person of all those persons having the surname e.g. Smith, is referred to).[8] From this we can conclude that the proper name is in se no logical operator for individuation, i.e. for distinguishing an individual from all other individuals that exist, though this is often believed. Yet, in all societies the need exists to refer to persons in an unequivocal way, but, as we shall presently see, this can be done in a variety of ways.

What is different in naming persons in primitive thought compared to our own customs is that the total system of social relationships reflects the religious order sanctioning it and lending it a timeless character, something that existed in the past in our culture too, but has been lost a long time ago. This made possible an alternative to characterising uniquely an individual by a proper name: at a definite moment, he or she could be *the unique member of a class*, this class being an element of the total traditional order under consideration. Indeed, if the structure of the conceptual system is thus that it contains a number of "places" or "positions" for persons, to be the only one who occupies a definite place in the social system means to be uniquely characterised, not by his or her own characteristics, but by those borrowed from the place that is filled in. However, this place in the conceptual system is not in its turn defined by the peculiar individual that occupies it momentarily; it remains unchanged under a succession of individuals that occupy it.

Cl. Lévi-Strauss sees in the often complicated social structure, which is a hierarchy of layers of pairs of opposites, arranged symmetrically or asymmetrically, in pairs or in groups of three or more, a means to come to terms with the reality of everyday life, its difficulties and tensions. As an example we can refer to name-giving practiced by Osage and Omaha Indians. For a child to be assigned to one subgroup or to the other is extremely important, and there are strict rules to be followed, such as alternation between the two subgroups according to order of birth. The role the individual will have to play, his character and temperament are settled beforehand by his mode of affiliation to the father.[9]

> This modelling of the person following specific schemes that are elementary or categorical does not only have physical, but also psychical consequences. A society that defines its segments in function of what is high and what is low, of heaven and earth, of day and night, can cover in the same structure of pairs of contraries ways of being social or moral, conciliation and aggression, peace and war, justice and order keeping , good and bad, order and disorder, etc. From this follows that it does not limit itself to the contemplation in the abstract of a system of correspondences; it offers a pretext to the individual members of these segments to singularise themselves by their conduct; and sometimes it even incites them to do so ....[10]

The receiving of a name is part of the process of integration in one of the clans. The name is often composed of three elements: a name indicating that one belongs to a definite clan having its own totem, a name indicating the position inside the clan and then a third name, the core name, that is worn only by one person or by one person at a time. It is clear from this that the composite name signifies localisation in a totality, and that, at least in a restricted community, this localisation ultimately is a perfect way to individuate a person. An example of this localisation in a totality can be seen in Fig. 2.6.

Today we seldom reflect upon the Christian names and surnames we wear and they have become meaningless. Nevertheless, they are mostly those of saints and still in the previous century by wearing such a name a person was supposed to be under the protection of the saint in question. When we look closer at surnames we can see that once, when people lived in small communities, these names must have contained useful information. Indeed, examples abound indicating that they referred to a local reality: Evans, Johnson, Jones, Harrolds, Peters(on) Simps(on), Thompson (lineage), Mac Donald, Mac Leod (clan), Whitehead, Goodman (physical or moral characteristics), Fowls, Fox, Wolf, Wren (animal), Church, Hall, Blackwell, Brook, Bush, Wood, Westman, North, Moore (reference to a place or a wind direction), Clark, Cook, Baker, Bishop, Dean, Draper, Marshall, Miller, Smith, Steward, Tanner (profession), Black, Brown, Gray, Green (colour), Fleming (nationality), ...

In the totemic system either each thing and everything has meaning or nothing has.[11] However, names can be used in a more equivocal way, and their exact meaning can be deliberately concealed for outsiders. For example, in Sioux tribes the names of the clans can be an aspect of something, e.g., Cakwyamtiwa, literally "blue that has appeared," but meaning in fact, following the clan, burgeoning flower of the tobacco plant, or burgeoning flower of *Delphinium Scaposum,* or the germination of plants in general. It seems that the general aspect of a series of phenomena is grasped or, what comes to the same, the identity in the diversity, and then used as a circumlocution. This way of naming is necessary, in order to avoid that certain names of plants or animals serving as eponyms for individuals that are taboo for some religious reason, have to be pronounced.[12] This is a cultural element that is not specific for totemism alone. Belief in reincarnation as such can be a reason for kindred prohibition and goes together with vegetarianism for obvious reasons.

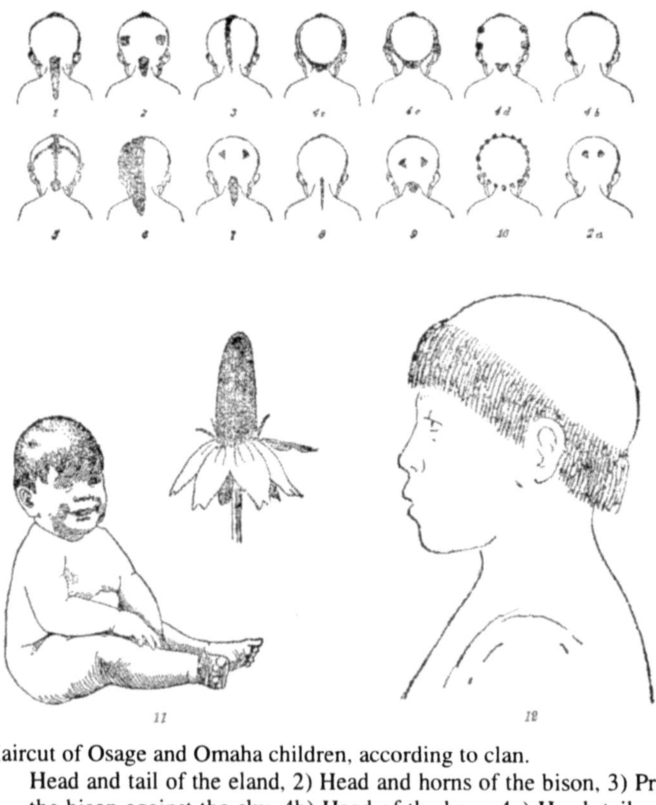

Haircut of Osage and Omaha children, according to clan.

Head and tail of the eland, 2) Head and horns of the bison, 3) Profile of the spine of the bison against the sky, 4b) Head of the bear, 4c) Head, tail and body of the little birds, 4d) Shell of tortoise with head, legs and tail,4e) Head, wing, tail of eagle, 5) Four points of the compass, 6) Hairy flank of the wolf, 7) Horns and tail of the bison, 8)Head and tail of the deer, 9) Head, tail, growing horns of the young bison, 10) Teeth of the reptile, 11) Rock surrounded by floating algae, 12) Corn flower.

Fig. 2.4. *Ibidem*, 226.

The names of a clan are its property and privilege; they are jealously guarded. A clan possessing many names is considered rich, a clan disposing of a limited number of names poor. In some tribes guardians of the names are appointed, who at each moment know exactly which names are available, i.e., which ones are not worn by anybody. When a child is born, the guardian is called for and he can inform the parents which names can be given. Sometimes, more children are born than the number of available names, and children may have to wait for years, until a member of their family dies.[13] Cl. Lévi-Strauss reports that such strict rules can be observed not only in Indian tribes, but also in Africa.

The Dogons of Sudan follow a very strict method for attributing proper names, which consist of locating the position of each individual following a genealogical and mythical model in which each name is linked to a sex, to a lineage, an order of birth, and to the qualitative structure of the group of closely related persons to which the individual belongs.[14] Furthermore: "By its rules and customs each society applies a rigid and discontinuous grid on the continuous flux of generations on which it imposes a structure."[15] Different ways exist of achieving this. Individuals can be given an *autonym*, a name without reference to somebody else; or they can be on the contrary referred to by an *eponym*, a name relating a person to another person, such as the name of a hero, or they can be referred to by a *teknonym*, namely "father of...," "mother of...."; they can be referred to by a *necronym*, indicating that he or she is somebody whose father or mother is dead, or by a *patronymic*, "son of...," "daughter of...."

In systems based upon *classes of relations* names are used relating persons with other persons, but in order that the stability of the structure is secured, all reference to proper names is avoided. A person is simply called "grandfather" or "grandson." For the relation between persons to become truly a class term the proper names must disappear from the expression used to refer to an individual, because proper names as an integral part of the name would deprive the expression of its generality.[16] The other system of name-giving implies *classes of positions* in the system, a kind of name giving we considered already before. It is more stable, though it has the disadvantage to be less adaptable to the rise and decline of the population. Here the term contains a proper name, but is used as a class with only one member at a time. Indeed, if there are more candidates for a name than free places, it is possible that one or more individuals temporarily cannot be satisfactorily identified by a name.[17] What has become clear is that in this type of thought the individual is considered *both* as a *kind* of person and as a *unique*, but replaceable person, a substitute. In Europe, in the Middle Ages, as long as enough Christian names were available to avoid confusion in small communities where people all knew each other, one name was sufficient for recognition, mostly the name given during the ceremony of baptism. This ceremony, during which a new member was added to the community and identified as such by its baptismal name, was an important element in the process of Christianization. From the eleventh century on there seems to have been an insufficient number of available names, and there was a tendency to add a second element to the first name, referring to a family relation, a profession, a personal characteristic, a place, etc. Only royalty and nobles could afford to use simply the Christian name of their father, but sometimes a numeral index was added, or a topographical element.[18] In the past, and this till recently, certain traditions were respected, and names contained a minimum of meaning. Not so long ago the first-born son received his Christian name from his paternal grandfather. Often the name of the godfather and /or of the godmother was added. Cl. Lévi-Strauss draws our attention to the fact in France and in Belgium, until recently, it was a custom that women, after the death of their husband, simply put "widow of" before the name of their late husband, rather than returning to their maiden name.[19] Nowadays, we can only derive from a name to which family somebody belongs, and this not even in an unequivocal way, as many families that are unrelated have the same name. Our names are rather poor means of identification having no further reference to an actual situation or conspicuous facts. In our individualistic society most of us are anonymous in spite of having an official name, save in a very narrow circle.

History

A last element that can contribute to the comprehension of the self-conception of primitive men is the "a-historical" character of their cultures. It would be a mistake, as Lévi-Strauss has explained in a brochure *Race et Histoire*, which he wrote for UNESCO, to consider the history of the human race to be comparable to a ladder, each step of it meaning a progress. Undoubtedly, cultures can become more and more refined and advance step by step, but these processes are local phenomena. Like the evolution of biological species, the overall development of cultures over time has rather to be represented like a bush than like a ladder. We think of cultures in terms of what is of interest *to us*, and consider them as superior to other cultures if they have scored in domains *we* consider important. We have no good criteria to compare them, and yet we are irresistibly driven to order them hierarchically.[20] Another problem is that no longer existing pre-historical cultures, generally defined as cultures of which no written records exist, though they sometimes have left enough traces for a hypothetical reconstruction, seldom have provided us with enough clues for the serious study of their history.

Nevertheless, primitive people do have a history, of course. For example, they fight wars, and in past centuries, they have been decimated by epidemics; their numbers can decline or grow in function of a number of other factors, such as the expansion or, more commonly, the reduction of their territories. However, history has a very different meaning in this context from what we ordinarily understand by this term, not because they cannot write and read and consequently their traditions are not written down, but because their society reacts upon change in an *extremely conservative way*, and sudden changes in their social and cultural life have catastrophic effects. As far as possible they transform their social and cultural system, without damaging its main characteristics, but just enough to enable them to cope with the new situation. We have indicated the reason for this: their social and cultural order reflects an eternal world order and must continue to do so. Moreover, because the individual members of the social group consider themselves to be elements taking up given positions in the overall structure, they are bound by such a tight system of rules, that the concept of free will and self-consciousness, so dear to those who believe man is a truly historical being, has here no meaning. The Salesian fathers understood the importance of the circular form of the plan of the village of the Bororo and of the orientation of the huts in relation to the wind directions for the religious life of these people. In an effort to convert them to Christianity they made them leave their villages and settle in houses built in parallel rows; thereby their individual and social life lost its meaning and they became human wrecks.[21]

Marxists and existentialists alike have criticised the theories of Claude Lévi-Strauss, because he clearly exalted the culture of primitive men. He believed that, their approach to reality being probably the original one once shared by all men, its structure can be recognised in the most fundamental layers of the cognition not only of men in the past, but also at least in part of his contemporaries. In a way, it could be said that the relation of so called "savages" with nature and their mentality, according to Lévi-Strauss still present in us, was to be preferred to the cool rationality and practicality of modern man. In this thought analysis into elements and sub-elements is complementary to the opposite movement, the reconstitution of the totality giving meaning to its parts. In primitive thought the sharp distinction we make between nature and culture, science and religion, object and subject, rationality and emotion, does not exist, or rather, the opposite poles are arranged in such a way that they are in balance. Most important in this context is that contraries can be reconciled: *the general is also what is particular, the particular is also general.* The Marxists should have recognised the affinities of dialectical reasoning with savage logic, but they probably failed to do so. They were more interested in what they considered to be the political implications of Lévi-Strauss's analysis: they could not accept the a-historical and anti-revolutionary character of the society of primitive man. The Existentialists from their point of view could not accept that the "humanity of man" was not his self-consciousness, his pure ego and the fulfilment of his freedom. Both movements, Marxism and Existentialism, are offshoots of idealism, and it does not come as a surprise that they rejected this vision of primitive man as inauthentic and alienating. If their representatives had been consistent they should have drawn the conclusion that these primitive people were not men in the true sense of the word according to their definitions. Indeed, even if they would have been conscious of their "historical" situation, they were not ready to change their society drastically and to transform the world by the production of their own goods, not any more than they were pondering about their existence as their own free creation.

How we think about time and about history can throw light on the differences and similarities between primitive thought and our own conceptions. Primitive cultures are conservative; in this view what is appropriate is what the ancestors would have approved of. If it can be shown that ideas or behaviour is in accordance with the sacred traditions they are thereby justified. Our own culture has been very conservative in the past. (Let us not forget the famous battle in the late Middle Ages of the "Ancients" and the "Moderns." The idea that what was new could also be positive was not readily accepted, but eventually the mentality changed).

Claude Lévi-Strauss stresses that the primitive conception of time and history, if well-understood is not inferior to our own vision, but establishes a harmony between the diachronic dimension and the synchronic dimension of human reality, between what changes in the course of time and what remains the same and is outside time:

> No doubt it is difficult for us not to judge negatively an attitude that blatantly contradicts this avid need of change that is proper to our culture. However, the stubborn fidelity to the past conceived as an eternal model, rather than as a moment in becoming, does not betray a moral or intellectual shortcoming: it expresses a consciously or unconsciously adopted choice, of which the systematic character is attested, in the whole world, by this indefatigable repetition of the justification of each technique, each rule, and of each custom, by means of one argument: our ancestors have taught it. Like for us in other domains till recent times, antiquity and continuity are the foundation of legitimacy. But this antiquity is posited as absolute, as it goes back to the origin of the world, and this continuity does neither allow orientation, nor degree. Thus, mythic history shows the paradox of being simultaneously disjoint and conjoint in relation to the present. Disjoint, because the first ancestors were of another nature than contemporary humans: the former were creators, the latter were copyists; and conjoint, for, since the apparition of the ancestors, nothing happened, save events of which the recurrence periodically wipes out the particularity. We must still show how savage thought is capable, not only to surmount this double contradiction, but to extract from it the elements for a coherent system where a diachronic dimension that is in a way domesticated, collaborates with the synchronic dimension without the risk that between them new conflicts arise.[22]

Notes

1. *Ibidem*, 219.
2. *Ibidem*, 193–194.

3. *Ibidem*, 197.
4. *Ibidem*, 201
5. W.K.C. Guthrie, The Greek philosophers. From Thales to Aristotle, Routledge, London, New York, 1989, 101–102.
6. Lévi-Strauss, *La Pensée Sauvage*, 221.
7. *Ibidem*, 222.
8. *Ibidem*, 249.
9. *Ibidem*.
10. *Ibidem*, 227.
11. *Ibidem*, 228.
12. *Ibidem*, 233.
13. *Ibidem*, 250.
14. *Ibidem*, 251.
15. *Ibidem*, 264.
16. *Ibidem*, 262. Cf. also: *Genèse médievale de l'anthroponymie moderne, Ie et IIe Rencontres* 1986–1987, Azay-le-Ferron, s. d.
17. Lévi-Strauss, *La Pensée Sauvage*, 261.
18. Cf. *Genèse médievale de l'anthroponymie moderne*.
19. Lévi-Strauss, *La Pensée Sauvage*, 256.
20. Lévi-Strauss, *Race et Histoire*, Editions Gonthier, © UNESCO, Paris, 1961.
21. Lévi-Strauss, *Tristes Tropiques*, 189.
22. Lévi-Strauss, *La Pensée Sauvage*, 312–313.

Chapter 3. The conflict between primitive ways of thinking and Aristotelian logic

The continuation of primitive thought in Greek philosophy

Few people are aware of the fact that the primitive way of conceptualising the world has survived in historical times. The first Greek philosophers of the sixth century B. C. are renowned for having introduced a rational and naturalistic approach to reality. These men did not use myths or religious beliefs to explain the order in the world; they explained the phenomena they studied in terms of matter considered as an active principle. However, quite another type of philosophy flourished at the same time, which was a powerful combination of science and religion. Pythagoras of Samos was already famous when he arrived in Kroton. He founded a new mystery religion based upon Orphism, and could soon gain the confidence of the Krotonians. Numerous adepts entered the Brotherhood and accepted unconditionally the authority of their leader. Though the Brotherhood was short-lived, its influence upon Greek culture was immense. Two of the most famous philosophers of antiquity, Socrates and his pupil Plato, became adherents of Pythagoreanism. The philosophers of nature were interested in material processes, Pythagoras in the abstract form of natural phenomena and the spiritual aspect of reality.

We can wonder whether the merging in Pythagoreanism of religion and speculation with mathematical science was an advantage or on the contrary a disadvantage. Joseph Campbell is convinced that the mix of both was beneficial; he believes that mythical thought is a necessary component of all cultures, including our own with its extraordinary development of science. Arthur Koestler, who has studied the history of the development of a scientific conception of our solar system, the place of the Sun in it, the movements of the Earth and of the other planets of which it is composed, considers the philosophy of Pythagoras as the paradigm of a harmonious vision, the perfect balance between rationality and speculation. He writes:

> The essence and power of that vision lies in its all-embracing, unifying character; it unites religion and science, mathematics and music, medicine and cosmology, body, mind and spirit in an inspired and luminous synthesis.[1]

Cl. Lévi-Strauss has a view similar to that of Koestler: he is convinced that the ancient mentality, combining religious myths and knowledge, has many advantages over the cool reasoning and the scientific and technical approach of reality by contemporary Westerners. What Koestler admires above all is the equilibrium in the importance accorded to knowledge of natural phenomena and to abstract forms of thought. This just measure of both can no longer be found in the philosophy of Plato, who despised the phenomenal world and knowledge acquired through the senses, but at least religion and a rational world view were combined in his philosophy. In Plato's cosmology God is *in* the world that is a living being with a soul. Aristotle, by contrast was probably not a deeply religious man. He reduced God to a logical necessity, to an unmoved mover who merely set things in motion *from the outside* and then left the world to itself. Science is essentially the discovery of the causes of the natural phenomena. Here the so praised harmonious mix is lost. Nevertheless, seen in the context of his time, Aristotle was a great scientist and Plato was not. What I want to show is even more, to wit that science on the one hand and religious thought and "transcendent" metaphysics on the other are incompatible and this not only by their respective contents, but also by the logic they use.

In primitive thought, we have stressed that the objective and the subjective, rationality and belief, are mingled. Concepts forming pairs of opposites and belonging to very different domains are associated with each other and ordered into often intricate structures. To primitive people there is no strict separation between the categories of concepts which today we believe, after Ryle, should be clearly distinguished in order to avoid confusion and contradiction. It is clear from the following scheme that the Pythagoreans had no qualms whatsoever to use for interpreting reality pairs of opposite concepts belonging to very different domains. To them they were interconnected ontological principles referring to different aspects of one and the same reality. Let us consider the following scheme (Table 3.1.):

Ten pairs of opposites.

Limited	↔	Unlimited
Odd	↔	Even
One	↔	Multiple
Right	↔	Left
Male	↔	Female
In rest	↔	Moved
Straight	↔	Curved
Light	↔	Darkness
Good	↔	Bad
Square	↔	Oblong

Table. 3.1. Jean Tricot, (introduction, notes and translation), *Aristote, La Métaphysique*, Vrin, Paris, 1970, 2 vol., Vol. I, 45–46.

Among the various elements of the list are mathematical concepts that today are emotionally neutral for us, though in the past their mystical significance has played an important role in all sorts of speculative thought. As we can see, integrated in the scheme, the elements of the pairs are either on the good side or on the bad side and accordingly have a positive or negative meaning.

We must not think this speculative interpretation of mathematical concepts was proper to Pythagoreanism. We can see the link with primitive thought comparing the ten Pythagorean pairs of opposites that are essential for understanding reality with for example the mythical numerology of the Osage tribe, mentioned by Lévi-Strauss. In the central myth of these people the most fundamental contrast is that between two halves, heaven and earth, the latter divided into land and water. This is the starting point from which developed a complicated "grammar," a system of correspondences of concepts ranging from the concrete to the abstract domain. According to Lévi-Strauss:

> a logical structure—a simple pair of contraries to start with—expands into a sheaf in two directions, one abstract under the form of numerology, the other concrete, first elementary, then specific.... [2]

In the numerology of the Osage tribe the number 6 belongs to the heavenly half of the world, the number 7 to the earthly half, and their sum, 13, corresponds in their cosmology to the number of rays of the rising sun, which is *half* the sun. On the social level 13 corresponds to the number of valiant actions an accomplished warrior must be able to boast of, a warrior who is a *half* man, because the war function is the privilege of only one of the two halves of the totality of the tribe.[3] The initial scheme functions as a catalyst, triggering off the elaboration of other schemes, which are binary, ternary, quaternary or even of a still higher numerical order. We shall not enter into the details of the intricate cognitive structure Lévi-Strauss analyses here. What is of central interest for our subject is that here too the even is opposed to the odd, 6 to 7, the even being, contrary to the Pythagorean conception, associated with what is light and lofty, the odd with the earth.

The reason why the odd is on the side of the positive concepts and the even on the negative side in the Pythagorean list is the following. The Greeks abhorred what is infinite, because it has no boundaries. A square is beautiful, because for any length of its four sides, when the measure of a side is divided by the measure of another side the result is always 1, while the division of the long side of a rectangle by the short side is variable.

The mathematicians used a "gnomon" or carpenter's square to demonstrate certain properties of numbers and the correlated geometrical forms, as in Fig. 3.1.

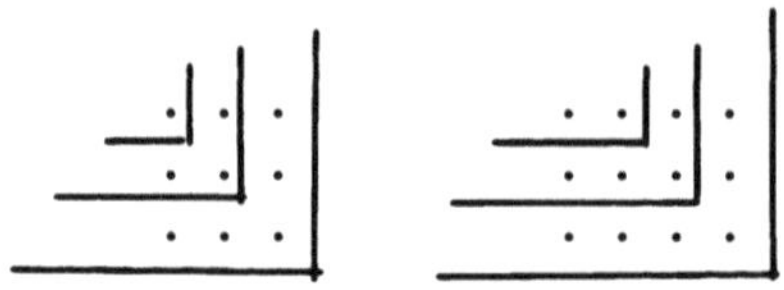

Fig. 3.1. Cf. Tricot, *Aristote, La Métaphysique*, Vol. I, 44–45.

It can be seen that with odd numbers in the gnomon we obtain squares, with even numbers rectangles, following the formulas:

$1+3+5\ldots(2n-1) = n^2$ $\qquad$ $2+4+6\ \ldots +2n = n\,(n+1)$

Plato, as we can read i.a. in his *Timaeus,* still clings to the traditional logic in which pairs of opposite concepts are basic. In se there is nothing wrong with contraries. Aristotle himself, who was less of a poet than his master, looking for the causes and constituents of nature, comes to the conclusion that being cannot be one, but that the four elements are its simplest manifestations, and he orders them in pairs of contraries according to their properties: Earth is dry and cold, Water cold and moist, Air moist and warm, Fire warm and dry. Though opposite concepts can be useful, and Aristotle used them without any qualms in his physics, as we shall see, he repeatedly declared that they cannot be the basic elements for a sound theory of valid reasoning.

In Western culture Aristotle's logic is fundamental. The examples of the Aristotelian system of classification known to all of us are generally taken from biology, but we could easily find similar schemes in all other scientific domains. Moreover, we use the same type of classification in daily life, for example to order our belongings. Normally we put all things that fall under the heading "clothing" together in a wardrobe. Most people put their shoes at the bottom of the wardrobe, and asked why they probably will answer that it is for hygienic reasons. They hang their clothes on clothes hangers on a bar, ordered according to their similarity, the blouses or shirts with the blouses or shirts, the vests with the vests, the skirts with the skirts, the trousers with the trousers, the coats with the coats, etc. In a separate compartment, on shelves or in a chest of drawers, they put their underwear, sorted in neat stacks, socks with socks, underpants with underpants, etc. To most Westerners a sock is a sock; it is a piece of clothing that protects their feet, keeps them warm and dry, and nothing else. We do not think about our socks, underpants, trousers, skirts, etc., as belonging to the realm of the lower half of the universe, and we could without causing a scandal put our shoes in an upper compartment of the wardrobe. It is not in contradiction with the world order and not sacrilege to put them next to hats. They simply have no special meaning, they are what they are. Compare from this point of view the ordering of the content of the wardrobe of a Westerner with that of an Amerindian as represented in Fig. 3.2.

The importance of this difference in our way of regimenting things and that of primitive people becomes also clear in the remark of Lévi-Strauss about the shocked reaction of members of a tribe of Amerindians, who saw white people, when making up a bed, inadvertently inverse the upper and the under sheet. To them, this contradicted the order of the world. All that exists is primarily divided into things that belong to that what is "upper" and all things that belong to that what is "under."[4] The examples of the primitive way of regimenting various elements of reality abound. Thus the confection of moccasins is highly ritualised in the tribe of the Osage, because, next to being utilitarian objects, they have also cultural a meaning. The moccasin is that with which the grass (= earth) is trod. Therefore, it belongs to the upper realm in so far as it crushes the "bad" grass and is opposed to it, but it also belongs to the lower realm because it comes down on the ground and touches it. The solution for this dilemma is the performance of rites to accompany the confection of shoes that reconcile their positive and negative aspect.[5]

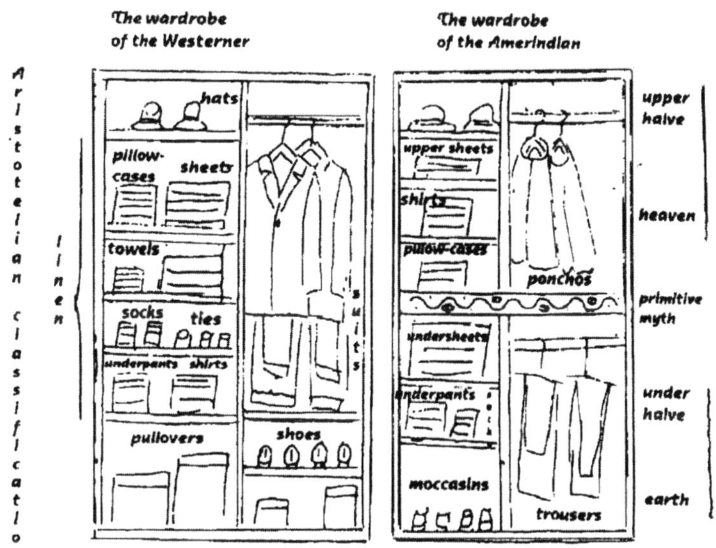

Fig.3.2 Imaginary wardrobe of a Westerner and that of a "primitive" Amerindian.

Another example of ordering according to opposite concepts is that of the Peul:

> The Peul of Africa, who are the Bororo of the Nigerian zone of the Sahara, seem to associate, like the Kaguru, the right side to men, and in the temporal order with what comes first, the left side with the women and with what comes afterwards; symmetrically, the male hierarchy runs from South to North, the female hierarchy form North to South. It follows from this that in the camp the woman arranges her gourds in an order of diminishing size, placing the biggest one in the South, while the man tethers his calves in the inverse order. (Dupire).[6]

What is crucial here is that the logical scheme in terms of contraries is applied in detail, not only in a religious context, but also in daily life. We can see from the case of the Peul that not only heaven and earth, but also right and left is an important distinction. The combinations of such pairs of concepts with one another can be endlessly varied. In the case of the Peul left and right are combined with spatial orientations: right with the south-north direction, left with the north-south direction. They are at the same time combined with temporal concepts, namely "before" and "after." In the scheme of the Pythagorean world order we have encountered this same contrast between right and left, right here being a positive concept, left a negative concept. The Romans inherited this idea and to them what came from the left (sinister), for example a flight of birds, was a bad omen. This preference for the right side that exists even today in our culture is a remnant of primitive ways of thinking that probably stems from the fact that most men are right-handed. The French call a clumsy person "gauche," Flemish people qualify such a person as "possessing two left hands."

We must not believe that incompatible logical systems necessarily mutually exclude each other in the mind of a person. Orthodox Jewish people all over the world order their kitchen following strict religious rules. They have a cupboard for crockery, pots and pans, knives, and other utensils that come into contact with milk products and with farinaceous products cooked with milk, while another cupboard contains the same utensils, but coming into contact with blood, such as meat, etc. The reason is that in the Old Testament it is said that "the lamb shall not be cooked in its mother's milk." Though orthodox Jews apply the rules of the Torah with rigour, they simply use the Aristotelian logic we are familiar with in all matters where no specific religious prescriptions are relevant, and they are not aware of a contradiction in their thought. The case is far from unique. We should expect that in highly developed cultures the old ways of thinking have disappeared. Nevertheless, in rural China, where Western cultural habits have had little influence, the ancient mentality persists, and the contrast between yin and yang is still today the central principle following which all things are classified and understood. Five elements reign and ideally keep each other in equilibrium; too much of one of them or too little and the harmony is lost. Though at the most fundamental level knowledge directly related to sensorial experience must be similar in all humans, and though there is something like a core of "common human interests," on a cultural and more sophisticated level variations are possible in conceptual ordering structures and the ways they are filled in.

Aristotelian logic compared with primitive logic

Classification according to primitive logic is based on opposites. Wanting to classify a specific variety of pepper for example, the primitive mind says what it is and what it is not: thus the pepper in question is something that can be named in contrast with what cannot; it is a thing and not a person, nor an animal; it is a plant and not a stone; it is a herb and not a ligneous plant; it is a pepper and not a rice plant..... (Cf. Fig. 2.2.). Primitive people are keen observers and in this example the result is not very different from a western botanical classification, but this is not always the case.

Aristotle's logic was a new system for ordering and regimenting the elements of nature and the basis for further reasoning. It was far superior to dialectical reasoning and classification by division or "*diairesis*," and it was generally applied not only throughout Antiquity, but it was also known and used right from the start in the Middle Ages. Let us compare the classification by division with the Aristotelian system.

The Aristotelian classification, instead of starting from *opposites*, starts from *similarities*. Reality consists primarily of first substances, i.e., concrete individual things. Accordingly, Aristotle's classification considers *individuals* to be the basic elements. Individual things have essential properties and accidental properties. Only if we know the essence (the sum of essential properties) of things and start from it can we classify them properly. Classification proceeds by considering those substances that share a number of essential properties. It is supposed that we have the knack to discover which things belong to one kind by a process of abstraction in which the accidental properties are discarded and those shared by all the individuals considered are retained. Individual things of one kind fall under the least general concept possible, the *species specialissima*. Together with other *species specialissimae*, which resemble each other, because they share common properties, it falls under a slightly more general concept, a species. This species, together with other species showing similarity, can be considered a genus, which together with other genera falls under a still broader concept, a class, which is again compared to other classes, etc.

The distinction between general and particular is clear-cut. First substances, i.e. those things that can exist independently, are concrete individuals and hence particular. Secondary substances, such as properties, never exist autonomously, but only in as far as they are inherent in first substances. Being inherent in *many* similar substances, they are general. The concepts used for classification, even the *species specialissimae*, are general. Therefore a species has no autonomous existence; it is not a substance. Individuals are irreducible; they can be grasped intuitively by an act of perception, but, all knowledge being general, they cannot really be known. Jean Tricot has, in a comment in his edition of *Metaphysics* of Aristotle, rightly observed that Aristotle's ontology and epistemology cannot be reconciled without difficulty, all that exists being particular, while all knowledge being general.[7] Also, in the Middle Ages *omne animal ineffabile est*, "each animal is ineffable," was a common dictum amongst scholars. The concrete, particular individual participates in a number of general concepts, and it can be described in general terms, but, as such, it is never completely characterised by them.

Aristotle himself classified in *Parts of Animals* animals grouping them according to the qualities they share. Modern commentators have gathered the scattered indications he gave and reconstructed his scheme. The first and basic distinction made by him is between animals with red blood (enaima) and without it (anaima). (Table 3.2.) The two groups have eight subdivisions.

These eight groups are called *great genera*, by Aristotle; their subdivisions are termed *genus*, and the subdivisions of the genera are termed *species:*

I Enaima:

1. Quadrupeds that are viviparous, to which belong all the Mammals, comprise also the cetaceans, the seals and the bats. This first group is once more divided on the basis of a second criterion, namely the skeleton and the form of the extremities.
2. The quadrupeds that are oviparous (lizards, tortoises, Amphibia).
3. The Birds, subdivided into eight kinds, according to their extremities (with claws, separated toes, palmated), and according to their diet (grains, insects, etc.).
4. The Fish, subdivided following the nature of their skeleton in cartilaginous and osteoid.

II Anaima:

This class comprises also four groups:

1. The anaima with a weak body (Cephalopods).
2. Those with a weak body covered in scales (Crustaceans).
3. Those with a weak body covered by a strong shell (shellfish, sea-urchin, ...)
4. Insects, subdivided into nine species. To them are added the worms.

Table 3.2. Cf. Aristotle, *Parts of Animals*, with transl. by A.L. Peck, Harvard Univ. Press, Cambridge, Mass., London, 1993, 22–23.

The advantages of his logic according to Aristotle

Aristotle was aware of the advantages of his logic compared to the logic based on opposites. On several occasions he expresses his views on the subject and explicitly rejects the method of Plato.[8] In order to classify things we need definitions based upon similarities: all substances of one kind share their essential properties. The definition is a combination of the genus under which the species falls directly and the specific difference between the genus and the species in question. It is clear that only the essence of a kind of thing permits to define it with logical necessity. This is how he illustrates this: when looking e.g. for a definition of man, to be stated in an affirmative proposition, the rule is that the middle term must be subordinate to the major term, and cannot be a universal that includes it. This yields the following correct syllogism:

"All animals are mortal. (It is an essential quality of animals to be mortal). Man is an animal. Therefore man is mortal."

The process of division, on the contrary, takes the universal as the middle term. The logic of contraries does not offer the possibility to *prove by inference*, it merely begs the question:

> As in the case of conclusions reached without the use of middle terms, if it is stated that, given certain conditions, such-and-such must follow, one is entitled to ask "Why?"; because the divider can state and prove (as he supposes) by his division that everything is either mortal or immortal; but such a proposition taken as a whole is not a definition. Thus even if the proposition could be proved by division, the definition still does not amount to a logical inference.[9]

Aristotle has not only explicitly rejected the method of reasoning by division based upon contraries and shown that not only it never results in good definitions,[10] but, and to him most important, that starting from pairs of opposites one cannot prove the fundamental *law of the excluded third*.

Socrates defined concepts using his dialectical method. He questioned his interlocutors about concrete examples, then he tried to establish which of these examples were accepted by his audience as falling under the concept under examination, and then attempted to find an agreement about what they had in common. This clarification of concepts leading to definitions was continued by Plato. However, the latter introduced separate realities, ideas or forms corresponding to the concepts. They were eternal, perfect, one and at the same time general, because there existed a vast number of copies of them, but these copies were individual, particular, limited in time, which for Plato meant that they were imperfect and only real in a lesser sense when compared to their models.

This dual conception of reality separates the individual from the general. As a consequence, in this respect the logic of mythical thought, where the particular and the general are aspects of the *same reality*, is here lost. Aristotle then takes the decisive step: he rejects the reality of eternal ideas and there is again only one reality. General concepts, universals, which are the product of a process of abstraction in our mind, take the place of transcendent ideas.

I have drawn attention to the fact that Aristotle's classificatory system was revolutionary. It starts from particular elements of reality, makes abstraction of their accidental properties, brings them on the basis of their common essential properties under one concept, the *species speciallissima*, makes a synthesis of similar concepts that fall under a more general concept, etc. The difference with Plato is clear. O. Hamelin, in his study on Aristotle's system, writes:

> Plato, with his method of division, anyway so inadequate for the task he intended it for, has had the suspicion we treat of here [that pairs of contraries are capital for a philosophy of concepts using the synthetic method], because the difference is opposite to the genus, and the species is a synthesis of both, and in consequence, it is without surprise that we find here, if not a general theory of the contraries, at least an already firm sketch of a solid and careful doctrine that has implications for the capital form of contraries called contradiction. The position of Aristotle is very different. He does not know any other rational link between the concepts than the analytic one. The principal reason there is to meditate about concepts that are each others contrary in consequence did not exist for him.[11]

We use Aristotelian logic in science and daily life

As we can easily see from modern systems of classification making use of Aristotelian logic, by discarding differences considered as unimportant and looking for resemblance by common properties, a conceptual scheme is obtained in which the least general entities are subsumed under a concept that, together with other similar concepts is subsumed under a broader concept, etc. An example from biological classification will make this clear:

Group of equiform classes: *phylum*	vertebrates: fish, reptiles, birds... *(chordata)*
Group of equiform orders: *class*	mammals: monkeys, rodents, ungulates (*Mammalia*)
Group of equiform families: *order*	animals of prey: felines, bears... (*Carnivora*)
Group of equiform genera: *family*	canines: hyena, dogs.... (*Canidae*)
Group of equiform species: *genus*	dogs: wolf, jackal... (*Cania*)
Group of equiform races: *species*	domestic dog (*Canis familiaris*)
Group of equiform individuals: *race, variety*	Dachshund, Alsatian, ...

Table. 3.3. Cf. G. Köller, F. Anders, *Zoölogie*, Het Spectrum, Aula, Antwerpen/Utrecht,1966, 26. (Original title: *Zoologie*, Carl Winter-Universitätsverlag, Heidelberg, 1963).

As we can see (Table 3.3) the Aristotelian system has no other ontological implications than the existence of individuals and of concepts that are the product of the activity of our mind. We can also represent the scheme of Aristotelian classification as a *genealogical* tree and introduce thereby the element time (Table 3.4.). (Aristotelian logic, however, is in principle timeless).

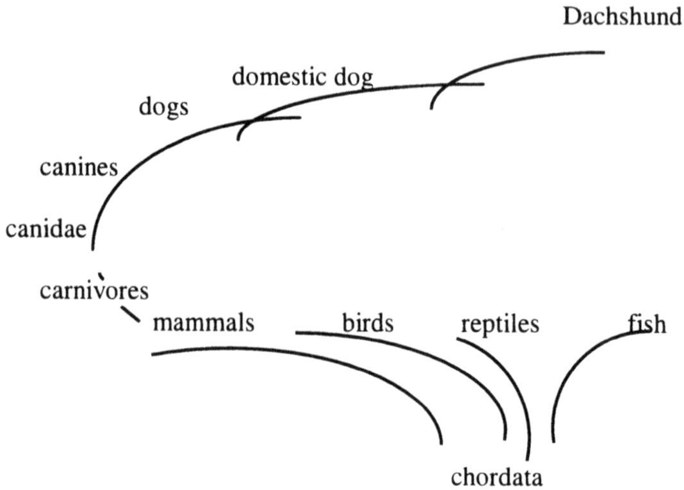

Table. 3.4.

The relation between metaphysics and religion on the one hand, science and logic on the other

Can science, metaphysics and religion form an integrated whole? This is a central question to which we seek an answer in this book, because in our culture this is no longer evident. If they are incompatible and contradict each other, this would have of course important consequences for our conception of man, which would thereby become double. This is indeed my thesis: Westerners of the twentieth and twenty-first-century still cherish a picture of man that has its roots in religion and metaphysics and at the same time they dispose of scientific knowledge that contradicts this picture.

What I want to treat in this section is the role of Aristotelian logic in the long cultural process that resulted in this situation. Could we say, that this specific form of logic, which is unique to our culture, has been a major factor in the disruption of the equilibrium between the religious sensibility and the irresistible drive to know, between the need to lend a general meaning to things and to know them objectively? It is certain that transcendent metaphysics and religious speculation, which both start from the certainty that the source of the world we experience is a reality behind or above it that is its foundation, has always rejected the hegemony of Aristotelian logic that was only useful in purely scientific and practical matters. I do not want to suggest that there is only one form of rationality, but that some forms of it are incompatible with Aristotelian logic, which is the logic of practice and of science in the West. Primitive thought, which was our main concern in the previous sections, is not pre-logical and not irrational, but it cannot be used for the purposes of science. We must be aware of the fact that we find in various Western philosophies mixtures of "pure" rationality with remnants of the primitive type of rationality, and that they either reserved for Aristotelian logic a restricted domain where it was appropriate or that they even totally rejected it.

A good example of a mixture of science and religious metaphysics is to be found in the work of Plato, which can be considered a paradigm case. Let us go a little deeper in this matter. Plato was fighting a rearguard action, by defending in a time when the city state was already doomed, the idea of such a state as an independent political, economic and social unit. The anthropomorphic gods of the Greek pantheon were worshipped at festivals organised by the city and it was a duty for the citizen to attend them, but not to believe in the gods.[12] Plato's aim was to oppose to the strong tendency in philosophy towards relativism, naturalism and atheism, a rational philosophical system based upon a theistic conception of the creation of the world order and the government of the universe.[13] This would enable him to provide laws and institutions with a transcendent, unquestionable principle on which to found them.

In *Timaeus*, he explains how the world came into being, when a Demiurge created the Cosmos out of the original Chaos. Because the text contains so many parts that seem to be more the product of fantasy than of reasoning, the question has been raised whether this was not rather a myth of creation than an account to be taken literally. It was doubted that Plato himself believed all of it, but it seems that he was serious about the general content of this work. A hallmark of Greek thought in classical antiquity was its rationality, but it is of importance to specify what is meant by this term. His philosophy was based on reasoning and he also used mathematics in his cosmology for explaining why the world is as it is and could not be otherwise.

J.C.B. Gosling in his study on Plato has given a learned exposition of the problem of the role of mathematics in his work. Plato believed mathematics to be most useful in all theoretical and practical sciences. In *The Republic* he represents schematically the different levels of knowledge by dividing a line first in two, the parts having a certain proportion, then these two lines had to be divided again in the same proportion and he thus obtains four unequal lines. The lowest type is perceptual knowledge of what is only the mirroring of real things such as a reflection in water or the echo of a sound; slightly better is the direct perception of the things themselves; next, mathematics is real science, but starts from hypotheses that are not proven; finally, philosophy is the highest form of thought, because its aim is the intuition of ideas, i.e. to grasp them without further mediation. This means that, though mathematics is certainly the highest of all the sciences, it is surpassed by speculative thought. We know that Aristotle has said in his *Metaphysics*[14] that Plato considered the numbers to be a reality intermediate between the ideas and the phenomena. Nevertheless, the science of numbers and the corresponding geometrical figures cannot in itself reveal to us the ultimate truth about reality. According to Gosling, Plato started with the Socratic search for definitions, but he was not satisfied to obtain them by looking for an observable correlation between the phenomena. What he really wanted was not knowledge of the general and the unchangeable drawn from the empirical reality, but the general and unchangeable *in itself*, and this was just what according to him the Pythagoreans had succeeded in discovering. For example, the beauty of musical compositions is not due to the skill to combine notes and rhythms in an agreeable way, but to the application of invariable laws of proportion. Such mathematical laws can not only be found where music is concerned, but in all domains of the mind.

If this had not been only a first step in his evolution, Plato would have been satisfied to seek, like the Pythagoreans, merely the hidden mathematical framework that reveals the fundamental principles according to which all kinds of things are rationally constructed, and to escape in that way the uncertainties of the appearances. He wanted more: if the stamp of a rational making can be discovered in reality, then there must be a rational constructor at its origin. In the words of Gosling:

> Knowledge is of the good in that all knowledge is a grasp of intelligible principles, and understanding of the universe consists in knowing that it is constructed on such principles and in grasping the principles concerned. Being rational ourselves we appreciate such construction, and recognise its value, i.e. its rationality.[15]

The next step was to see that the creation was not only intellectually satisfying, but that this was the same as to discover in it what is good. Here we have left the domain of pure mathematical rationality.

In order to be able to appreciate the *value* of what is rational, more than mathematics is needed, namely the ability to appreciate the *right measure* in good and beautiful things. The knowledge of mathematical relations is a necessary condition, but it is not sufficient, it must be transcended. Indeed, as we have seen, mathematics starts from hypotheses that are not proven, whereas the ultimate knowledge starts form a principle beyond doubt, the cause of all things, or the *Supreme Good*. The idea of the Good is to be compared to the Sun; without its light nothing could grow and nothing could thrive.[16] In the same way the idea of the Good is the cause of all the other ideas, which are the eternal models, mirrored by the phenomenal world. In fact, if the universe is a coherent system realising the Good, then to know something is to know its relation to the totality and in this way to discover *what it is good for*. In this philosophy what is true and what is good coincide. (Also, in ethical matters, when a person acts badly it is because of ignorance.)

Plato does not seek material and efficient causes like the philosophers of nature, but an explanation based on reason or ground, and that is the way we must understand the word "cause" in his philosophy. This has of course implications for what he understands by "reason," which is a mixture of careful thought with intuition. In other words, reason is speculation, intellectual vision. (Gosling shows the strong metaphorical relation in Plato's epistemology between knowledge and visual perception).

For example, Plato explains that the Creator, in his goodness, gave to the universe the form of a sphere, because this is the most perfect geometrical form. He is not a jealous God and therefore he made it after his own image, giving it a soul that penetrates all its parts. The world is a closed system, without need to feed itself or to breathe, and is not subject to deterioration. After having established this, he devotes himself to the task of explaining by a dialectical mathematical reasoning why there are exactly four elements. God started making the world out of fire and earth, but it is impossible to make something that is good out of two things without a binding agent, and he used the best one, namely proportionality. This would have been sufficient if the universe had had to have only two dimensions, but because it had to be a body, two binding factors instead of one were necessary. Therefore God placed between fire and earth water and air and this according to the principle of proportionality.[17] Plato relates the four elements of which all things are composed to four of the five ideal solids: he associates earth with the cube, water with the icosahedron, air with the octahedron, and fire with the tetrahedron.[18] This type of philosophical reflection is *sui generis*, because it contains seemingly mathematical deductions that are in fact speculative. Indeed, though he does not refer to a pre-existing religious myth, he mingles his deductions with non-rational components. What belongs to the realm of the heavens is lofty, what is related to the earth and the senses is depreciated, what is circular or spherical is beautiful, good and superior to things showing other, less ideal forms. Thus the heavenly bodies are divine and therefore perfectly smooth spheres and they move in circles. Following the same principles the Demiurge, when he created man, gave to the mind a spherical housing, and in order that it should not be obliged to roll over the ground, he put it on a body provided with legs so that it could move about.

In his theory of the creation of living beings the contrast between high and low is fundamental. Plato inherited the theory of reincarnation from Pythagoras. People, when they are reborn, become animals of which the status corresponds to their merits. Accordingly, the soul of people who are interested in the path of the stars, but believe that they must look up at the night sky and use their senses to penetrate the heavenly secrets, does not return for all times to its home between the stars, but these people become birds after their death. Those men who are not interested in philosophical and astronomical matters, but are led by the passions in their heart, become land animals wandering about on all fours; moreover their heads are no longer round as in men but oblong. The more men indulge in being led by desire rather than by reason, the more they are drawn to the earth, and in consequence their reincarnation must be supported not by four but even more legs. Next come men who are even more blunted, and who become reptiles after their death, like the snake, without any feet, and not separated from the earth by legs. They are obliged to have direct contact with the ground and to slither about on their tummies. The most wretched men of all, delivered to their vile lusts, become animals that do not live on the earth, but in the lowest places, under water. They are not thought worthy by their makers to breathe air, but are forced to inhale muddy water and become fish or molluscs.[19]

Let us return for a moment to Plato's logic and dialectical reasoning. As we saw already, for discovering what a thing really is and defining it Plato uses the method of division. An example will be useful to understand this. Some divisions are adequate, some are not. Compare:

	man			man	
	=			=	
Barbarian (non-Greek)	or	Greek	male	or	female

The first one divides the concept by using accidental properties, of which the first is in fact negative, namely non-Greek, and is therefore inadequate. The second division is adequate, because here two positive properties are involved that belong to all men.

Let us now suppose that we want to know the content of the concept "politics" for example.

	politics	
	=	
	science	
	=	
practical knowledge	or	theoretical knowledge

politics falls under practical knowledge

	practical knowledge	
	=	
prescriptive	or	judgment

politics falls under prescriptive practical knowledge

In the previous section we have observed that Aristotle rejects the method of division, because it can never lead to a good definition. Indeed, as part of the dialectical method of seeking an answer by asking questions, this method is based upon the *opinion* of the persons who are interrogated.

In dialectical reasoning as it is conceived by Plato, the complement of the art to divide is the art to combine. Indeed, reality consists of elements that are combinations or mixtures. A mixture is always composed of an element that is indeterminate or unlimited and an element that is determinate. Indeterminate is a couple of opposite concepts that are relative. For example, greater and smaller, higher pitch and lower pitch, warmer and colder. The determinate element is a constant numerical relation: double, triple, etc. The mixture is the result of the introduction of such a constant relation in a couple of opposites. For example, musicians know that a relation of ½ introduced in the indeterminate couple of opposites, the "dyad" of high pitch and low pitch, creates the octave. In the same way, the introduction of a fixed relation in the couple of opposites slow and rapid creates a regular movement. The science of music must consist of the knowledge of the number and nature of the fixed relations. In his later work Plato did no longer exclusively consider binary relations; in the domain of music the number of terms of the relations was determined by the number of possible chords.[20]

Whereas Christian religion and Platonism have been successfully combined, religion and Aristotelian logic have from early on clashed. Thus the Apostle Paul, when preaching in Athens, had no impact upon his listeners, i.a. because the well-to-do citizens forming his audience were educated in philosophy and had at least some training in Aristotelian logic. How could a God who is eternal die? How could a man born from a woman be God? How could A be A and not-A? The Greek demand to understand could not be combined with having confidence and simply believe. Paul knew Greek philosophy, but he condemned it in the name of a New Wisdom that is madness and reason: the belief in Jesus Christ:

> The Jews demand miracles and the Greek are in search of wisdom; we, we preach a crucified Christ, scandal to the Jews and madness for the Gentiles, but for those who are called, be it Jews or Greeks, it is the absolute power of God and it is the wisdom of God. Indeed, the madness of God is wiser than the wisdom of men, and the weakness of God stronger than the strength of men. (Cor. I, 1: 22–25) [21]

This conflict between reason and intuitive belief will have repercussions in the Middle Ages, and this all the more so, because Paul had repeated more than once that logical reasoning was a weapon that could undermine faith.[22] It was not science that was the enemy of faith. It was logic, the application of its laws, resulting in incontestable necessary truths. The scholars interested in science borrowed the theories of the ancients that became available piece by piece in Latin translation and if necessary they interpreted them following the Christian tradition. In the eleventh and twelfth century, dialectics, not in the original sense used by Plato as seeking the truth by discussions, but reasoning following Aristotelian rules, was practised by many philosophers. As long as they used this logic as a means to give extra support to the Christian doctrine, this was accepted by the authorities of the Church, but when it was practised for its own sake, as *l'art pour l'art*, and applied to these truths without the intention of proving them, the authorities became aware that dialectics was a threat to orthodoxy and the culpable logicians were condemned. Roscelinus of Compiègne and Pierre Abelard only two famous examples. Roscelinus was a nominalist, and he believed that general concepts or universals are nothing but words. (Probably his theory was less crude than that, but we know it only as it was represented by his opponents). Anyway, to these words correspond those things to which they refer and no other reality. Thus a "wood" is a general term, to which corresponds nothing but a number of trees, and no reality apart from them. In se this is a harmless philosophical thesis, but consequently applied to the dogma of the Holy Trinity this leads to the conclusion that it consists of three separate particular elements and that their sum does not have an autonomous reality. Roscelinus was condemned and if he did not retract his contention, he had to be put to death by stoning, a very painful way to die, but he had the good sense to recant his philosophical belief. Pierre Abelard, pupil of Roscelinus, but also of William of Champeaux, a realist, took a more moderate position in the discussion about universals. He wrote a famous book, *Sic et Non* (Yes and No), about contradictory opinions of Christian authorities and applied logical reasoning to Christian doctrine. This brought him into trouble. Bernard of Clairvaux, who was often consulted as an arbiter in difficult questions concerning orthodoxy, declared about him: "When Abelard speaks about the Trinity, he talks like Arius, when he reasons about Divine Grace, he does it like Pelagius, and when he comments upon the person of Christ, he resembles Nestorius." In other words, "in this one man are present three heretics." He was condemned at Soissons and this made an end to his brilliant career as Magister.[23]

In the fourteenth century, William of Ockham, a nominalist and excellent logician, started from Aristotle, but stripped his logic of its metaphysical implications, namely essentialism and teleology. He referred to the work of Duns Scotus in claiming that faith and reason, science and theology, each had their proper domain. In spite of the fact that his philosophy was condemned by the Church, Ockham has been very influential; his ideas were passed on over the centuries, they were revolutionary seen in the context of his time and they cleared the path for future scientific progress. The nominalists paved the way for an empirical approach to natural phenomena, eventually replacing the old method that was based upon authorities rather than upon direct observations and logical reasoning. In Modern Times, slowly and after a long struggle, science finally relied upon the nominalist version of the logic of Aristotle. It was eventually admitted that essences, formal and final causes, which had hampered scientific progress for many centuries, had no explanatory value. Now the qualitative approach to reality could slowly be replaced by a quantitative one. I believe that we must see in the fact that Aristotelian logic was known and used not only in Antiquity, but throughout the Middle Ages, an important cause for the later development of modern science in Europe.

In the nineteenth century, science finally divorced from religion and metaphysics; scientific theories should not be based on extra-scientific convictions. Nevertheless, idealist philosophers still believed in the necessity of a "first philosophy" or metaphysics as the foundation of science or even as its precondition. Though traditional logic was perhaps suitable for practical matters it was unsuitable for speculative thought with which science had to be in line.

Kant proved by logical reasoning that metaphysics was impossible, because it could not provide valid knowledge. In imitation of Aristotle, he distinguished dialectics, which he considered to be illusory reasoning and based on a fictive logic, from sound Aristotelian logic. The transcendental dialectics he proposes is the study of the different forms dialectics can take and its aim is the detection of fallacious reasoning. Hegel was the main critic of the new practically oriented knowledge of which he considered Kant to be the main proponent. In his eyes the empirical approach of reality and therefore of man was an impoverishment. The study of the historical progress of human consciousness, in his opinion the source of all truth, had dried up. He bitterly regretted this evolution, and he blamed, at least in part, Aristotle for it. Science, in the true sense of the word, is metaphysics:

> The esoteric doctrine of the philosophy of Kant, —that our *powers of reason must not rise above our experience*; otherwise our capacity of comprehension would become *theoretical reason*, which does nothing but create *phantasms*—has justified for the men of science to abandon speculative thought. This popular doctrine meets the claim of the modern educational theory, the demand of the time, which has only eyes for the practical needs, the claim that like for knowledge experience is primary; for the good order in public and private life theoretical comprehension would be harmful even, and practice and practical education would be what is essential and the only thing necessary.—When in this way science and common sense cooperated to cause the downfall of metaphysics, then it seemed that the remarkable spectacle was brought about of an *educated nation without metaphysics*—like a for the rest manifold ornamented temple without a Holy of Holies.—Theology, which in former times was the guardian of the speculative mysteries and metaphysics, which it is true depended on it, had given up this science for the practical-popular and learned historical [knowledge]....[24]

The logic of daily life, but also of science and rational philosophy à la Kant, he says, is that of Aristotle, as we can see from the handbooks in use. It has neither deteriorated, nor progressed and is in great need of renewal. After two thousand years of evolution of human consciousness the logic of Aristotle is ill-adapted to the progress mankind has made. Also, he writes in his *Wissenschaft der Logik* (Science of Logic) the following lines:

> Kant for that matter praises logic, namely the aggregate of definitions and propositions, that we call logic in the common sense, being happy that it has early on been completed before other sciences; since Aristotle it has often consist more in omissions—it would be appropriate to conclude that it needs all the more a total reworking; because two thousand years of steady working of the mind has to have gained it a higher consciousness
> of its thought and of its pure intrinsic essence.... [25]

Traditional logic is in Hegel's opinion just good enough to give the impression that it can be used to learn how to think correctly; it is a series of formal principles that is only of use in the empirical sciences. More strongly, empirical sciences have their own method of classification and definition, and their logic is according to him itself a form of empirical science.[26] Theoretical thought, however, in the sense of pure subjectivity reflecting upon itself, is abstract; it does not need sense data to give it content, *as it is its own content.* Now that metaphysics went out of fashion, Hegel wanted to restore speculative thought to its former glory. In his opinion, reason is in truth spirit, which is higher than intelligent reason and than rational intelligence.[27]

The logic of metaphysics is dialectics, of which we can see a concrete application in *Phänomenologie des Geistes* (Phenomenology of Spirit):

> Here are forms of consciousness of which each one in its very realisation dissolves itself, and results in its own negation,—and is thereby turned into a higher form. The only way *to gain scientific progress,*—and the very *simple* insight we must try to reach,—is the acceptance of the logical proposition that the negative is also positive, or that what contradicts itself does not dissolve itself into nought, in abstract nothingness, but actually only in negation of its *specific* content, or that such a negation is not all negation, but *the negation of the specific thing* that dissolves itself, and that therefore it is a specific negation, and that in this way in the result that from which it results is essentially contained;— which is in fact a tautology, because otherwise it would be something that is immediately given, not a result. Because what is resulting, namely negation, is a *specific* negation, it has a *content.* It is a new concept, but a higher, more rich concept than the former one; because it has been enriched with its negation or its contrary, and contains it therefore, but also more than it, because it is the unity of itself and its opposite.—Along this way the system of concepts must build itself,—and must accomplish itself in a constant and pure movement that takes nothing from the outside.[28]

Like in primitive thought there is no neat separation of *subject and object.* Following Hegel *they are two sides of one thing*, because in dialectics the content does not come from outside, but the subject coincides with its content, and this fact makes it move. Thus, in his work on logic, Hegel first treats of Being, the most general and thereby the most empty of concepts, about which nothing can be said and which has no properties. Being is the thesis, of which Nothingness is the contrary and the antithesis. Being and Nothingness taken together produce Becoming, which is the synthesis of the thesis and the antithesis. Indeed, the negation of the specific content, *which it does not have, but is*, does not destroy it, but sets it as it were free to become another concept, another content.

This dialectics is based upon *opposites*, just like the logic of mythical thought, and this is the second point it has in common with it. The most important difference, however, is that it is not static, but *dynamic*. The synthesis of two concepts can be in its turn a thesis to which can be opposed an antithesis and both can be resolved into a synthesis that contains them under a new form. Mythical systems evolve by transformations, retaining their fundamental characteristics, while here thought is essentially a progressive movement towards an end driven by dialectics. It is indeed a principle of the mind rather than a mere method: concepts *produce* concepts.

The third analogy with mythical thought is that in dialectics there is *no absolute distinction between general and particular*; the particular is always "absorbed" in the general. For example: the individual can be considered the contrary of the community, because the interests of the individual are the opposite of the interests of society. Thesis, *the individual*, and antithesis, *society*, yield a higher concept by their synthesis, *the state*, in which the interests of the individual who has become a citizen are reconciled with those of the community. In this logic, nothing is ever unequivocally itself, nothing completely coincides with itself. The individual "produces" its contrary, because it "contains" in a sense the general, in the same way the general "contains" already its contrary, the individual. However, the general is primary compared to the particular. The human individual is only human to the extent that he is a concrete instance of the general concept "man." In fact the humanity of a human being lies in his participation in the general idea of mankind. Hegel declares that it is generally accepted that:

> If it is in this matter like it has been indicated before, and what is apart from what is generally recognised, namely that the *nature*, the real *essence*, what is truly remaining and substantial in the plurality and randomness of what appears to us and in its transitory expression, the *concept* of the thing, *is what is in itself general*. Like each human individual, though being something infinitely unique, has the *prius* [the first reason] of its uniqueness in being a man, like each animal has its *prius* in being an animal, so it would be impossible to say, what such an individual would still be, if this foundation would be taken away from it, even would it be equipped with however many other predicates, and though it [its nature] can be named just like the other predicates.[29]

The fourth element of resemblance between Hegel's logic and the logic of mythical thought is that in both *A is never A, but always A,' something that it is and is not.* Thereby the principle of identity and of the excluded third is explicitly rejected. Traditional logic is formal, Hegel says, because in it the concepts are distinguished from their content. The study of categories that are only formal is relegated to scholarly logic and metaphysics. Hegel's ironical comment is:

> The simple basic definition or common formal definition of the set of contradiction, is posited in the logic of this set. Healthy reason has to such a degree lost its reverence for this school, which is in possession of the laws of truth and in which these laws are still in use, that it laughs about it, and that a person who is able to say in truth: a plant is a—plant, science is—science, *and etc. infinitely many times*, is insufferable.[30]

The same thing can be said of the rules of deduction, which, though it would be unfair to deny that they are useful in the domain of (scientific) knowledge where they have their place, can produce untrue propositions and sophisms. In whichever way one defines the truth, for *the higher truth that is religious in nature* they are useless, because they only pertain to the validity of conclusions, to knowledge, not to truth as such.[31] In dialectical thought no concept is ever identical with itself, and therefore it is that concept and at the same time also another concept, otherwise it could not be productive. This is reflected in the existence of human beings: possessing consciousness and self-consciousness, they distance themselves from themselves, and thereby continually become something that they are not.

The idealistic tradition can be said to have largely determined Western non-scientific anthropological conceptions. It was continued in existentialism, which has been of great and lasting influence on the continent. We can find the same rejection of the logic of daily life and of science we encountered in Hegel's philosophy, in the work of most existentialist philosophers. It is prominent in the work of Heidegger, for example. In his *Was ist Metaphysik* (What is Metaphysics), in order to explain what the real purpose of speculative thought should be, he refers to Descartes's tree of knowledge: "Thus, Philosophy is like a tree of which the roots are Metaphysics, the trunk is Physics and the branches that spring from this trunk are all the other sciences...."[32] He rejects this conception of classical metaphysics; we must not look for the roots of all knowledge, but dig into the earth to discover the primitive ground (*Urgrund*) by which these roots are fed. This is a metaphor for his quest for transcendent reality, for Being, which goes beyond the metaphysical speculation about separate beings. He wants to pass from metaphysics to reflection about the Truth of Being, and this requires that we get rid of the dominance of logic.

Inquiring what is the domain of science, he comes to the conclusion that scientists say: "beings," *ta onta* in Greek, i.e. things, *and nothing else*. This "nothing else" is for him the occasion to introduce "nothing" and "nothingness," first as intensions of the mind: they are "objects of thought" and thereby not nothing. Parmenides, who is considered to have been the first logician, said: "Being is and Nothingness is not." According to Heidegger he was wrong: Nothingness *is*. The negation, "not," and "nothing," is not the origin of Nothingness, which is not derived from the negation; on the contrary, only the existence of Nothingness makes the negation and the concept "nothing" possible. This leads to the conclusion that the famous statement of Parmenides, "Being is and Nothingness is not," here becomes: "Being is and Nothingness is," which is in contradiction with the principle of the excluded third, at least if we accept that Aristotelian logic is the highest authority.[33] In the latter case the question would be self-destructive and could not be asked, the conclusion would have to be: "Nothingness is the negation of the *totality* of beings and therefore it is nothing, it does not exist."

Heidegger, like Husserl, believes that thought is never empty, but is always thought of something and if to think of nothing is impossible it would be a contradiction to think Nothingness. Nevertheless, as we have seen, for Heidegger Nothingness is more fundamental than the negation, and it is *founded in human existence*. Human beings, like all other creatures exist in the sense that they "are there," but they also exist in another sense by which they transcend the immediately given: they question themselves as part of the *totality* of all being, or as part of Being with a capital letter. From the moment Being loses its self-evidence, they ask: "Why is there Being rather than Nothing?," but they could not ask this question if Non-Being did not exist, because then the question would not make sense. Both are inseparable: Nothingness pervades everything, because no being is self-evident. The circle is closed. The human being is the only one of all beings that can make himself questionable. Therefore in this fundamental questioning lies his humanity. Nothingness is given; it is present in our existence as something that we can experience at certain moments, for example in the joy the nearness of a beloved person gives us. He or she is there, but could also not have existed. In this experience not only the miracle of the other person is revealed, but Being reveals itself. We are presented to ourselves with a question mark; we "open up ourselves" and no longer simply coincide with ourselves, and take ourselves and all the other beings for granted. Yet this questioning of Being does never really end. Being can be compared with the horizon, we can go into its direction, but never reach it. Instead of *ex nihilo nihil fit* (from nothing comes nothing) of the Ancients, Heidegger comes to the conclusion: *ex nihilo omne ens qua ens fit* (from nothing comes all beings qua beings)[34] And his comment upon Hegel's "Pure Being and pure Non-Being is therefore the same thing" is:

> Hegel is right in formulating this proposition. Being and Non-Being go together, but not because—from the point of view of Hegel's conception of thought they correspond to each other by their indeterminacy and immediacy, but because Being itself is in essence finite and because it re-

> veals itself only in the transcendence of the existence that is immersed [literally: "held out"] into Non-Being.[35]

Thus Heidegger has shown that the dominance of logic in science, which is also present in traditional, i.e., immanent metaphysics, must be broken if we want to return to speculative thought of a more primitive kind, as, according to him, it once existed in the teaching of some pre-Socratic sages.

The conception of man in Hegel's and in Heidegger's philosophy is fundamentally different from that of Aristotle, who considered a human being as a substance, ontologically irreducible in his particularity. For Hegel the humanity of man is that as a spiritual being he can transcend his concrete materiality by being related to the general and ideal. For Heidegger the humanity of man is his capacity to take himself no longer for granted, but to seek the ground of his being in a transcendent reality. Neither philosopher is interested in the biological reality of human beings. What is important for them is the uniqueness of human existence and the sense it makes. Their conception of man corresponds to the deeply felt need to ascertain that to be human is more than to be a rational animal and that there is more to existence than "to hatch, to match and to dispatch." Most people cannot accept that their existence has only the sense they give to it, but that is not intrinsic in it. Their question is "is that all there is to it?" There *must* be a reason why we exist, there *must* be something transcending material reality that makes life ultimately worth while. We cannot find it by reasoning according to Aristotelian principles, but instead we must rely on the peculiar logic of speculative and/or religious thought.

Notes

1. A. Koestler, *The Sleepwalkers*, Penguin, Pelican, Harmondsworth, 1986, 27.
2. Cl. Lévi-Strauss, *La Pensée Sauvage*, Plon, Paris, 1962, 193.
3. *Ibidem*, 189.
4. Ibidem, 190.
5. *Ibidem*, 188.
6. *Ibidem*, 191.
7. Tricot, *Aristote, La Metaphysique*, Vol I, 439–442 .
8. René Taton, La science antique et medieval, Des origines à 1450, Quadrige, PUF,1994, Vol. 1, 273.
9. The Platonic method of dichotomy: *Cf. Plato. The Collected Dialogues*, Bollingen Series LXXI, Princeton University Press, Princeton, 1999, *Sophist*, 219 A ff., *Politicus*, 258 ff.

Criticism: Aristotle XVIII, *Metaphysics, books X–XIV*, Harvard University Press, Cambridge Massachusetts, 1990, Λ, 10, 1075a, 25 ff. ; *Aristotle, Prior Analytics,* Harvard University Press, Cambridge Massachusetts, 1973, I, XXXI, (358–361), and *Aristotle, Posterior Analytics,* Harvard University Press, Cambridge, 1976, II, V, (188–191).

10. *Aristotle, Posterior Analytics*, II, V, (191–193).

11. O. Hamelin, *Le Système d'Aristote*, edited by L. Robin, Félix Alcan, Paris, 1931, 128–129.
12. W.K.C. Guthrie, *The Greek Philosophers Thales to Aristotle*, Routledge, London and New York, 1989, 82–83.
13. *Ibidem*, 88.
14. J. Tricot, *Aristote, La Métaphysique*, Vol. I, p. 60–62, A 6, 987, b, 20–25.
15. J.C.B. Gosling, *Plato*, Routledge and Kegan Paul, London, 1983, 101.
16. Cf. *Plato. The Collected Dialogues*, Bollingen Series LXXI, Princeton University Press, Princeton, 1999, *Republic*, VI, 508, 743.
17. *Ibidem*, *Timaeus*, 31–32, 1163–1164.
18. *Ibidem*, 55–56, 1181–1182.
19. *Ibidem*, 91–92, 1210–1211.
20. Cf. E. Bréhier, *Histoire de la Philosophie*, Presses Universitaires de France, Paris, 1951, Volume I, 133–135.
21. Cited by Gilson in: E. Gilson, *La Philosophie au Moyen-Age*, Payot, Paris, 1962, 12.
22. *Ibidem*.
23. G. Leff, *Middeleeuwse Wijsbegeerte*, Prisma, Aula, Utrecht/Antwerpen, 1967, 120. (Orig. title: *Medieval Thought. St Augustine to Ockham*, published by Penguin, Harmondsworth, 1961.)
24. G.W.F. Hegel, *Wissenschaft der Logik*, (Vorrede zur ersten Ausgabe), Felix Meiner, Hamburg, 1963, 3–4.
25. *Ibidem*, (Einleitung), 32–33.
26. *Ibidem*, 34–35.
27. *Ibidem*, (Vorrede zur ersten Ausgabe), 6.
28. *Ibidem*, (Einleitung), 35–36.
29. *Ibidem*, (Vorrede zur zweiten Ausgabe), 15–16.
30. *Ibidem*, 17.
31. *Ibidem*, 17–18.
32. M. Heidegger, *Was ist Metaphysik*, Vittorio Klostermann, Frankfurt, A.M., 1949, 7.
33. *Ibidem*, 25–26.
34. *Ibidem*, 36.
35. *Ibidem*.

Part II.

Philosophers on the Subject of Man. A Concise Historical Overview

General Introduction

About Western culture and the sources of its conception of man

Till the fifteenth century the most important cultures of the world can be said to have reached comparable levels of development. The culture of the Chinese was as refined and sophisticated as that of the Europeans and their science is generally considered to have been even in advance of that of the Renaissance on the European continent. The same holds for Arab culture, which was in many respects by no means inferior to ours, but which on the contrary influenced Europeans during the Middle Ages in several domains. Above all it transmitted to us the works of the ancient Greeks. In the Middle Ages scientific knowledge in Europe amounted to an amalgam of what was retrieved from the legacy of antiquity, and no spectacular progress was made. From Modern Times on the situation changed; the most conspicuous distinctive characteristic of Western culture became its unique scientific development, which from then on gained momentum. From the occupation of individuals disposing of a sufficient income to have the leisure to study and experiment, science became in the eighteenth century a matter of national interest, and scientific societies and academies flourished in several countries. Accordingly the prestige of science grew enormously; progress in this domain is now seen as a source not only of national pride, but of cultural and economical advancement. The following step is that science is no longer a matter of personal interest, but becomes an enterprise. Universities, and soon thereafter scientific foundations and industrial laboratories, offer the possibility to pursue a scientific career. Science is prestigious and becomes a power in society to be reckoned with, so much so that those who try to ignore it are accused of obscurantism.

It is a curious fact that, though science divorced in the nineteenth century from religion and metaphysics, even then the conception of man did not change for the majority of Westerners, at least not in its most fundamental aspects. Darwin's theory of evolution, though notoriously in contradiction with most important dogmas of Christianity, did not shake enough our traditional beliefs about ourselves to reject them completely. The naturalistic view of man offered by science did by no means triumph over the idea that we have a unique ontological status, and even today it can be said that the general Western conception of man remains a product of the almost millenary religious and metaphysical tradition.

The constitutive elements at the origin of Western culture are Christianity and Greek philosophy, mainly Plato and Aristotle. In the Middle Ages science consisted by and large of what was retrieved from the legacy of antiquity, and it cannot be said, as mentioned above, that it, progressed in a spectacular way. Also, the general picture of man was based upon religious truths combined with the Christian version of Platonism and later on the Christian version of Aristotle. Reason and faith often clashed where theoretical matters were concerned; however, it must be noted that reason must not be understood as scientific knowledge, but as "dialectics," reasoning following the logic of Aristotle. When this logic was applied to matters of faith, this often led to unorthodox conclusions. The natural sciences consisted chiefly of a concatenation of citations of ancient authorities the content of which was quite harmless, and scientific theories played hardly any part in the opposition between rationality and faith. Discussions of the nature of man were mainly theological. When finally modern science developed this did of course change our view of the world, but not in any fundamental way our conception of ourselves.

Reasons for the development of science

There have been many speculations about the causes of the development of modern science in the West. Some, like I. Wallerstein, have thought that they were *economical*, that the unique progress made in Europe was mainly due to the rise of *capitalism* that originated in the peculiar economical and political situation at the end of the Middle Ages on the European continent. Others have stressed that the cause was mainly to be sought in the peculiar nature of the Judaeo-Christian *monotheistic religion*, in which man was created after the image of God and ordained by God to rule over Nature. This seems to be contradicted by the fact that Islam is a monotheism sharing with Judaeo-Christianity the basic pattern of its world view, which is derived from the Old Testament, and that the Islamic cultures did nevertheless not produce a modern science comparable to that of the West. However, it is certainly true that in Western culture man and nature, the soul and the body, subject and object were opposite poles. Man is not part of nature but has the task to dominate it. This led in Europe to the idea that knowledge is power not only over nature, but that it procures also political, social and economical power. This was probably an important stimulant for the development of science.

It can be argued that the extraordinary growth of Western science is to be sought in our love for novelty, which is quite remarkable compared to the conservatism in other civilisations. It leads to the *progressive disruption of the chief intellectual traditions*. The willingness to abandon them, to research independently and to bring innovations was a precondition for scientific advancement. In Europe the intellectual legacy crumbled and split into numerous fragments, represented by a multitude of different schools, spiritual movements and churches combating each other. The question is now why we are such a querulous lot. Perhaps part of the answer must be found in the political situation. At the end of the Middle Ages and in the Renaissance, politically our continent was still a vast patchwork of principalities, a feudal system. It was a world economy because it was broader than the existing political units, but not an empire with a central political power keeping the units together. The unifying factor was that we were all Christians, but the oneness of the Church was challenged at the end of the Middle Ages, when a schism arose and there was a Pope in Rome and another one was installed in Avignon. The unity was completely lost after the rise of Protestantism. This was a crucial moment in the cultural history of the West, because a general view of orthodoxy no longer existed.

Religion was certainly a state affair, but in each state separately, as no central power was present, strong enough to impose one religion. The reason for Chinese traditionalism, e.g., was probably that China was an enormous country, and the primary interest of the Emperors was to administrate it properly and to guard its immense borders. The stability of the empire was better served by a conservative hierarchy of civil servants educated with respect of the traditions than by innovators. Conventional knowledge was safeguarded by the mandarins who for their appointment had to pass examinations in literature, philosophy and religion. Traditional knowledge was passed on again and again; little interest existed for innovations in the human sciences or for improvements in the natural sciences and for further technological progress. Though the Chinese certainly had the means and the knowledge necessary for the exploration of unknown parts of the world, they considered China as a world in itself, having no need for expansion, which would imply foreign influences. In Europe the density of the population created a chronic lack of space, and its various kingdoms were always keen to expand their territories. This can explain why the countries of the Iberian Peninsula, Spain and Portugal, which had no such possibility of expansion because of their geographical position, were interested in the exploration of unknown parts of the world.[1]

Greek thought, which is considered to have been the cradle of our culture, had a dual character right from the start. Its legacy is on the one hand a naturalistic representation of reality, on the other hand a metaphysical and religiously inspired world view. The former had hardly any influence on European culture, whereas the second has put an indelible stamp on it. Especially Plato and Aristotle have been of major importance, the former by his religious and metaphysical views, the latter by his philosophy of course, but also by his logic and his scientific work. The importance of this logic for the development of modern science can hardly be overrated. It was a substantial improvement on dialectical reasoning, being not only an instrument for valid reasoning, but making *definition, sound classification and demonstration* possible.

After the fall of the Roman Empire much of Hellenistic culture was lost in the Western part of Europe, but remnants of it were safeguarded by the learned monks of the early Middle Ages and passed on by them from generation to generation. Latin being the language of the educated people, the Latin authors constituted the link with the ancient Greek. Boethius was the main source for the knowledge of the logic of Aristotle. As E. Gilson writes:

> The work of Boethius is multiform, and there is not one of its aspects that did not influence the Middle Ages, but nowhere was its authority more widely spread than in the domain of logic.

Then follows an extensive enumeration of his works on the subject and Gilson continues:

> One could say, that by the entirety of his works Boethius has become the professor of logic of the Middle Ages till the moment that, in the XIIIth century, the complete *Organon* of Aristotle himself (i.e. the totality of his works on logic) was translated in Latin and directly provided with comments....[2]

This situation did not change in later periods; his logic was taught without discontinuation. At the end of the Middle Ages William of Ockham rejected its usual realistic interpretation and gave an account of it that had no longer ontological implications. Essences and natures were mere concepts to which no reality separate from their individual instances corresponded, qualities, existed only in concrete particular things, not in se. In the following centuries the quantitative approach to reality replaced step by step the qualitative approach. Aristotelian logic was seen as universal, valid in all domains, save where the major Christian dogmas were concerned.

As an example to illustrate the contrast with other cultures we can refer again to China. Contrary to Westerners the Chinese remained faithful to the old, primitive logic and this for millennia. Speculating about the cosmos they constructed a conceptual system in which all the elements were linked with all the others, following the principle of polarity.

Fig. 4.1.

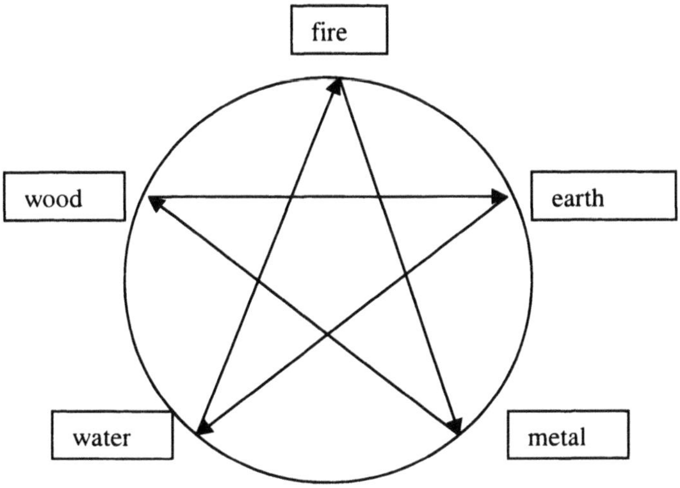

Fig. 4.2.

In nature two principles reign: Yang is male, active, life-giving and light and Yin is female, passive, receiving and dark. Yang belongs to the realm of Heaven and Yin to that of the Earth; they are respectively represented by a white and a black surface united in a circle. Each of the surfaces contains a dot in the other colour symbolising that something of the opposite element is always present in it. Nothing is totally what it is, but always also its opposite, which is the fundament of the harmony of the world. (See Fig.4.1.) There are five elements and they can be ordered upon a circle into pairs of contraries. (See Fig. 4.2.) This basic conceptual framework is not reserved for the speculative domain, but is applied in all matters.

Though China made important efforts to adopt Western scientific theories and technology from the nineteenth century on and this after having for millennia believed in its own superiority, even today in this vast country the old mentality and sanctioned ways of thinking persist. Successively decried in the new republic after the fall of the Mantsjou dynasty, suppressed under the government of Tsjang Kai Tsjek, first forbidden under the communist regime because it was not in accordance with "scientific Marxism," then again in favour under Mao, because being profoundly "dialectical," the traditional conceptual approach of reality has survived even today. In many domains of the Chinese culture it is still predominant. For example, though in the cities Western medicine is taught and practised, it is much too expensive for the vast majority of Chinese, who prefer to recur to traditional medicine. Acupuncture, which is best known in the

West, is only one discipline of Chinese medical science that is governed by the principles of primitive logic. It combines harmoniously empirical knowledge and world view. Man being a micro-cosmos reflecting the macro-cosmos, the approach is holistic; ailments are the consequence of an imbalance in the "Qi" or flow of energy in the individual that causes a disturbance in the equilibrium of the different functions of the organs, rather than that ailments are the consequence of a failure of these organs separately. Indeed, individual things are not isolated from each other, but they are rather the expression of the working of the two principles that are contrary and yet striving for balance, alternating, but also completing and determining, destroying and creating each other.[3]

When in Europe the prestige of science grew, it was seen as a threat to religion by Catholic and Protestant authorities alike. Science, from its side, over time had developed a methodology that no longer allowed that religious or metaphysical views were taken into account in scientific theories. Religion, metaphysics and science estranged from each other and became separate domains. The consequences of this are nowhere as important as where our self-conception is concerned. After having characterised the self-conception of men in primitive cultures I want to present a picture of how in the West the general non-scientific view of man grew historically.

Notes

1. Cf. I. Wallerstein, *The Modern World-system, Capitalist Agriculture and the Origins of the European World-economy in the Sixteenth Century*, Academic Press, New York, London, 1974.
2. E. Gilson, *La Philosophie au Moyen Age*, Payot, Paris, 1962, 139.
3. Th. P. Van Baaren, *Van Maansikkel tot Rijzende Zon*, De Haan, Standaard Boekhandel, Antwerpen, s. d.,127–128. Cf. also P. Mole, *Acupuncture*, Element Books, Shaftesbury, 1992, *Acupunctuur*, Kosmos, Utrecht/Antwerpen, 117–118.

Chapter 4. Antiquity

Philosophers of nature: man a natural phenomenon

The religious background

Let us start with the cultural background of the birth of philosophy in ancient Greece in the sixth century before Christ. In the Mediterranean region Babylonian, Egyptian, Hebrew, and, of course, Greek religion thrived at that time. All of them, except Judaism, are monophysisms; they imply namely that there is only one reality. Nature comprehends everything, the sky and the earth with its rocks, mountains, rivers, but also living beings, plants, animals, men, spirits and Gods that are identified with forces or elements of nature and can be worshipped in them. The God of the Jews was originally also associated with certain impressive natural phenomena of which he became probably a kind of synthesis. It has taken a long time before in the different Jewish tribes, of which each one adored their local Gods, monotheism was firmly established. What is special is that this one God, Yahweh, did not coincide with Nature, but was a different kind of reality altogether: he was what we would call *transcendent.* This means that he could not be perceived by the senses, though he could if he wanted *reveal* himself to man in nature and did so repeatedly. The Hebrews called the divine presence "el" and "elohim," a kind of force that could be experienced at certain special places and moments, not directly, but by a token, or in a dream.

It must be noted that Greek religion too was exceptional because, in contrast with the other religions flourishing in this period, it was not based upon holy texts, but upon a literary tradition, namely the works of Homer. Therefore, the Greeks, though probably as religious a people as others, had a different conception of faith and religious duty; it did not consist of a kind of orthodox credo, but of the observance of the rites that was identical with showing the Gods respect. Moreover, the distance between them and their Divinities was smaller than in other religions, because the Greeks believed they were the distant relatives of the Gods. However slight, a family resemblance was present. Moreover, some mortal individuals had inherited a divine spark that enabled them to be "God-like." They were a kind of transitional beings, the half-Gods, like Hercules or Prometheus. The contrast with Hebrew religion is extreme in this respect. Yahweh to Hebrews is the Creator, the source of all that exists. The Greeks believe the world is eternal, a Chaos out of which a Cosmos was born. The first philosophers were going to ask how the ordered world came into being and this is our next subject.

Ionian culture

The Ionians possessed a large degree of spiritual freedom. The fact however that the Ionian philosophers replaced for the first time mythical cosmologies by rational theories of nature had yet other reasons, which were to be found in the particular political and economical situation in Ionia, where the first philosophers lived. As early as the ninth century B.C., the Greeks started to colonise different parts of the Mediterranean. In the seventh and sixth century they had established commercial colonies in Asia Minor, amongst them Ionia. B. Farrington, a specialist in the field of the history of Greek science, pointed out that the pioneer spirit of the first generations of colonists, who were not interested in the possession of land, but in the establishment of trade routes over sea and commercial relationships with as many countries as possible, created a cultural climate favourable to innovation. No central government imposed its rules and laws in the colonies, no state religion dominated the cultural scene. On the contrary, the governments were changing, the cultural climate was liberal and tolerant, and many cults of different religions thrived side by side. The commercial aristocracy in power did not only encourage trade, but also the exchanges of knowledge and technical inventions.[1] Farrington also stresses that, though slavery existed in Ionia, it was not yet important enough for society to depend totally upon it. Craftsmanship was still highly valued, whereas in the fifth century in the motherland slaves performed all the manual tasks, tasks which in consequence were despised. These Greek Pioneers had little prejudices, were open-minded and tended to believe what their eyes and ears tell them. According to Farrington, it must have been this spirit that encouraged the philosophers of the school of Milete in Ionia to construct rational models of the world based upon empirical observations, rather than to derive world views from myths and religion.

These first philosophers, and all Greek philosophers after them, believed the world was not created, but eternal. It was at first a Chaos, a disorder, but it developed into a Cosmos, an order. What they sought was how this order came about. Their various answers were all naturalistic; to them all had started with matter that was active and could transform itself. There were four kinds of it, the basic elements water, earth, air and fire.

The question was which of them was the "First Principle" that gave rise to the other ones and to the world as we can perceive it. For Thales it was water; it is said that he came to this conviction, because visiting Egypt he had seen the silting in the Nile delta, where "land is born from water." Water could evaporate and give rise to air, of which fire in its turn is a rarefied form. Following Anaximenes, a pupil and friend of Thales, it was not water, but air that was the First Principle, perhaps because he had seen the process of felting and knew of course that water evaporates and tried to explain the formation of the world by a process of condensation and rarefaction. (In the opinion of the Greeks of that time air was vapour, and in consequence it could condense and become water, while water could evaporate and become air). Though these models can remind us of the mythical cosmogonies that tell us that at the command of a God dry land, a hill, rose out of the water, here the process does not require a divine order; it is conceived as a natural phenomenon of which examples can be observed in daily life. In this profane view an answer is given to the "how?" question of science, rather than to the "why?" question of mythical thought. (Some philosophers of science today contest this distinction, but here we see it is useful). The elements of nature are not Gods, their transformations do not coincide with the whimsical decisions of benevolent or malicious divine beings, but their transformations are lawful, and therefore we can speak of causal theories of nature. Anaximander, another philosopher belonging to the school of Milete, proposed as first principle the "apeiron" (what is undetermined) to explain the world order. Though we commonly speak of the school of Milete, as we can see from the foregoing, different opinions succeeded each other and cosmological problems were open to discussion.

Moreover, these philosopher-scientists had abandoned the logic of primitive thought: water was no longer conceived as the opposite of fire, air the opposite of earth, e.g., but on the contrary the four elements were considered as different transformations of one thing, matter. The approach of these men was naturalistic and quantitative; they were interested in what we would term today objective knowledge. Thales, for example, was a mathematician and had invented a method for the calculation of the distance of a ship from the shore. It is believed that he predicted a solar eclipse, a feat that seems to have made him famous in his own time. Space and time were no longer elements of a religious order, but were considered with the eye of a land surveyor and the mind of an astronomer, they could both be measured with mathematical tools. Anaximander seems to have been renowned for the improvement of the sundial permitting not only the measuring of the hours of the day, but also to observe yearly variations of the orbit of the sun.

Another element that is most important for our subject is that some of these philosophers explained life by a foreshadowing of a theory of evolutionism. Water is generally believed to be life-giving, as can be seen when land is flooded and becomes teaming with life after the waters have withdrawn. According to Thales all animals, man included, have evolved from fishes. Anaximander was of the same opinion: as living beings contain moisture and dead animals dry out, water must be life-giving. He also believed that men had started as fish, but later, after land had been formed, they became land animals. Men simply could not have been the first living beings, as their young need too much time to develop and grow up to be able to survive independently. Empedocles, also a philosopher of nature, believed that in the beginning parts of animals, heads and arms and feet and bodies were formed, which assembled in the strangest ways, and of course only the best combinations survived. The continuity between the species, including our own could not be accepted by religious philosophers; also in Christian culture the Ionian philosophers had no place and they were only rediscovered in the eighteenth century.

Philosophy of nature, initiated by the Ionian school, was continued by Heraclitus, Empedocles, Anaxagoras, then by the atomists Leucippus and Democritus. In the Hellenistic period Epicurus incorporated the philosophy of nature of Democritus in his own system. Democritus and Epicurus were quite famous in their own time, but, nevertheless, this type of philosophy was of nearly no influence upon the conception of man as it was developed in the West. Even in antiquity it was not really popular, because it was "atheistic," and because it made neither concessions to superstition nor to wishful thinking, as it offered no hope for a life after death.

Christian authorities have always considered atomism as an extremely dangerous aberration, totally incompatible with Christian faith. Reducing everything, even the mind, to matter, it could not be reconciled with the survival of the soul after death. Moreover, Democritus' physical theory was mechanistic and deterministic; in such a world no free will was possible. Determinism went counter to the Christian conception that God's will is absolutely free, and that he can intervene in the world at all times exerting his absolute power. Determinism also excluded the free will of men, the precondition of moral behaviour that distinguishes them from the beasts. Epicurus was more interested in the elaboration of a moral philosophy and used Democritus physical theory first and foremost as a weapon in his war against superstition. The supreme good for a person is to become happy and this goal can be attained by reaching a state of peace of the mind equalling the absence of all pain. To enjoy the good things of life with moderation and to control the passions were the principal recommendations, and this required a decision of the free will, the possibility of a self-discipline necessary to avoid all that could disturb the peace of mind. In order to safeguard free will, Epicurus introduced chance in the deterministic system: the atoms of the initial Chaos did not, as in the original theory, collide according to mechanical laws, but by chance, and the order of things we witness is the result of a chance process.

It could seem that from a Christian point of view the version of atomism of Epicurus was less reprehensible, but once the deterministic mechanical system of Newton became widely accepted, this seemed less obvious. Indeed, Kant drew the attention to the danger for religion it contained, by reasoning as follows: there is no need to prove that all the natural phenomena are separately willed by God. If we can explain them by the mechanical laws that apply to all matter let us not hesitate to do so, because in this way we pay tribute to Divine Providence. God created matter as an active principle bestowed with all the qualities necessary to produce the world as we know it. A far greater danger is hidden in chance than in necessity, because if the world order is the result of mere chance, God, who is a rational principle, cannot be at its origin. It must be said, however, that Kant was a deist, who considered the Bible as a collection of tales and parables. His God was not the God of the Bible, but a transcendent principle guaranteeing the world order. Therefore, Kant did not have to leave open the possibility of his free interventions in the world. (However, it was not without danger to criticise explanations of natural phenomena by Divine Providence). Orthodox Christians have always preferred physical theories containing an element of chance over determinism, because thus God's free will is not restricted by the laws of nature he has ordained, and in addition more room is left to the free will of human beings.

We now must turn our attention to the most renowned opponents of the philosophers of nature, Pythagoras, Socrates and Plato.

Pythagoras. A new type of philosophy

In the sixth century a new form of philosophy was initiated by Pythagoras. Though he was a native of Samos, an isle belonging to the Ionian sphere of influence, as we shall see, he was not a philosopher of nature after the manner of the school of Milete. He left Samos because of a disagreement with the local tyrant, and went to southern Italy where the Greeks had founded colonies, like in Ionia. There he started a new life, reforming Orphism that was in its turn a reformation of the cult of Dionysus. The legend of Orpheus tells us that the hero, after having failed to bring back his wife Eurydice from Hades, returned to the world. He longed for his wife and despised the charms of the Thracian women. The latter took revenge by tearing him to pieces under the influence of Bacchic frenzy. The parts of his body were buried, but his head was thrown in the river Hebrus. Carried to the sea floating on the water it kept singing. His lyre was placed among the stars. Pythagoreanism, like Orphism was a secret cult. Those who wanted to be members of the religious community had to go through rites of initiation, and after this the religious secrets could be revealed to them.

Pythagoras' philosophy was a curious mixture of religious elements and mathematical rationality. His aim was in the first place the salvation of the soul and mathematics was an exercise in spirituality. Like for the other early philosophers, we do not know much about him with certainty, and we must rely upon second-hand information. This has as a consequence that we often cannot make out what must be attributed to the master and what to his pupils and followers. Generally accepted is that Pythagoras believed that observable reality has a mathematical structure. (In a later version of his doctrine it is believed that the things we perceive in reality literally consist of mathematical points considered to be substances).[2] Probably the fact that he was at the head of a secret society explains that so little is known for certain about him. The idea of Pythagoras that the backbone of reality is its mathematical structure has been attractive to many people. In nature many forms can be discovered that can be described in mathematical terms, and it is easy to see in these forms the mark of God's hand. Anyway, we are warned by specialists of the philosophy of ancient Greece that we must not project our radical distinction between the material and the spiritual side of reality upon the philosophers of the sixth and fifth century B.C.; to them this distinction remained rather vague. When discussing arithmetical and geometrical problems, they drew with a stick dots and lines in the sand. These points had a surface, however small, the lines a definite breadth, and they were therefore at once material and spiritual. Represented in this way, numbers coincided with geometrical forms (Fig. 4.3.).

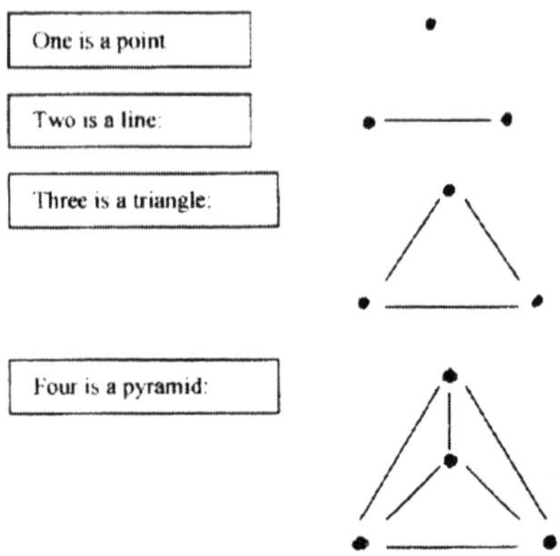

Mathematical relations were derived by Pythagoras from observations, at least, if we can believe the sources informing us about his life and work. Thus Boethius tells us that, hearing a smith hammering in his forge, Pythagoras was curious about the different sounds the man produced. He went in and asked if he could inspect the hammers and he weighed them. This gave him the idea of a mathematical relation between the pitch of the sounds and in this case the weights of the hammers making them, an idea that could be readily transposed to the sounds produced by the strings of different lengths of e.g. a harp. This first tentative solution of the problem of the harmony of sounds became known as the "tuning of Pythagoras." He was the first Greek philosopher to mingle rational and mystic elements in his philosophy. In contrast with the philosophers of nature, the question asked by the Pythagoreans is no longer "how?," but "why?," and the answers are often based on arbitrary associations. Though sensorial experience is not rejected, it is combined with beliefs and superstition. This is clear from the primitive answer why things are as they are that can be found by rational speculation. The latter can be seen from the symbolism of the numbers, which are thought to represent the essence of things: for example, 7 is occasion, 4 justice, 3 wedding.[3] The source of reality is "the one" that is at once even and odd and gives rise to the numbers in imitation of which the phenomenal world is made. The main pairs of contraries he used to make sense of the world strongly resemble the conceptual schemes found in primitive thought. Each of these pairs has a positive and a negative pole. Thus the first principle of reality is to be found in the opposition between limited and unlimited. Limited is good and beautiful, and when it is imposed upon the unlimited the result is harmony, just as when in the indefinite gradation between high and low pitch precise mathematical proportions are introduced, harmonic sounds are produced. From indeterminate Chaos determinate Cosmos is born. The universe is a living, breathing organism, a harmony of the male limited and the female unlimited. It is one in nature and this explains why transmigration of the soul is possible.

Conception of man

This leads us to the Pythagorean view of human life. The aim of Pythagoras was in the first place the purification of the soul. The members of the Pythagorean brotherhood had, like in many mystery cults, to undergo rites of initiation and had to turn their mind towards a spiritual reality and to resist the temptations of the flesh in the hope to escape the cycle of reincarnations and to be reunited with the universal soul of which once they were part. Longing for the redemption of their soul, they became capable of transcending their primitive condition by being conscious of the limitations of sensual pleasures. The believers could escape rebirth, but only after having reached the highest degree of perfection. In order to attain this goal they had to observe strictly the many rules of Pythagoreanism and a number of taboos. As said, our knowledge of this religion is secondhand and we do not always have a clear understanding of its prescriptions. One of their principles was vegetarianism. For example, they had to abstain from eating the meat of sheep, because, for those who believe in reincarnation, it is always possible that an animal is inhabited by the soul of a forebear. Other rules, such as the prohibition to eat beans for example, are less transparent.

The foregoing makes clear the difference with the more primitive cult of the fertility god Dionysus, in which during the festivals not only the drinking of wine, but also the ritual devouring of raw flesh was an important part of the orgiastic rites. Maenads, when at the height of their drunken rapture, killed young animals, tore their bodies apart, and ate their flesh. Dionysus had once been cut to pieces and thrown in a cauldron, but was resurrected by his grandmother Rhea, and the cycle of birth and rebirth in nature is thus symbolised. Pythagoras worshipped Apollo, the god of reason, and it is no wonder that in Pythagoreanism the instinctive element in human life is no longer honoured as a positive force, but rejected as contrary to rationality.[4]

Farrington stresses that Pythagoreanism announces a new mentality, the shift in interest for the natural phenomena to that for the realm of abstract thought:

> Mathematical relations replace physical processes or their states, like rarefaction, condensation and tension. The world, in the conception of the Pythagoreans, could be better and quicker understood by drawing diagrams in the sand than by thinking about phenomena like beaches having emerged above the water line, the silting of river mouths, evaporation, felting, and so forth.... Henceforth the banausic associations of the oven, the soldering iron, the bellow and the potter's wheel reduce their influence upon Greek thought in comparison with the more gentlemanly pursuit of theory of numbers and a geometry.[5]

Pythagoreanism survived, mainly through the work of Plato and through neo-Platonism, and it entered the mainstream of European thought. The brethren of the Pythagorean cult believed that it is by abstract speculation that a human being realises his nature in the best way. All things participated in the divine principle of the world and this was the path they had to follow to come near it. However, by stressing the unique rational faculties of man, they unknowingly paved the way for the anthropocentric view of reality that became prevalent in our culture.

Plato. An idealistic conception of the world

Pythagoreanism was a turning point in Greek philosophy: the philosophers of nature had lost their hegemony. Like his teacher Socrates, Plato was an adept of it. As is later on attested by Aristotle, Socrates rejected the study of nature and preferred that of man and of "the good life." Though Plato has written a cosmology, his main interest was not the natural phenomena, but the world of the eternal ideas, the ideal models of all things and values. The most perfect form of knowledge was in his opinion the direct apprehension of ideas by speculative thought. Mathematics, though the first of the sciences, made use of hypotheses and therefore was not direct, but mediated knowledge and therefore lower in rank. What the senses tell us cannot be trusted; Plato despised them, as much as he despised materiality. In the Timaeus, a work of immense influence in the Middle Ages, he applies his dualistic conception to the creation of the world, in which the rational precedes and forms matter. He explains how a divine being, the Demiurge or Divine Craftsman, ordered the original Chaos and made out of it the Cosmos in accordance with the Logos, i. e. with the laws of rationality. In his *a priori* model of the world Plato, is not like the Ionians looking for physical laws explaining how natural phenomena come about, but for the deeper reason why things are as they are. Like P. Duhem said, in his *Le Système du Monde*, for Plato the world is caused by an organised thought operating in function of a goal. In this teleological perspective, the universe results from the fertilisation of matter by ideas, while the ideas themselves are submitted to the ordering force of the divine thought or Logos.

Plato's God is rational and good. In consequence he chooses to make the best possible world, and because the sphere is the perfect mathematical body, this must be the form of the world. Movement in a circle being the most perfect possible movement the celestial bodies turn eternally round their own centre and in different directions round each other.[6] The Cosmos, in order to be eternal, must be a closed system, because if nothing exists outside it, no external influences can corrupt it. It is a living being, an organism made after the image of the God and therefore it cannot but be complete in itself and perfect. It is one and unique; no second world, and certainly not an infinite number of worlds, exist besides it.[7] Like all living beings, the world possesses a soul that is placed exactly in its centre and from there pervades it in all directions.

The creation of Man and the animal species

After having completed the description of the general order in the world, Plato tells us how the Demiurge creates the Gods, who in their turn have the task to create mortal men, their souls, their bodies. The Gods formed the body of men in accordance with its finality. The head, the most elevated part, is like the heavens a sphere; it must reign over the body. The latter, of oblong form, is made in such a way that it can move, and in order that it would not have to roll over the ground, limbs grew out of the body. The front of the head is moulded into a face, the seat of the senses, of which the eyes are the most important. Out of the transparent eyes streams the internal fire that is present in men and, during the day, the pure fire that is the light of the sun meets in our eyes a fire that is kindred with it.[8] Plato also explains how the different kinds of living beings came about; they were reincarnations of men of different moral value. This accounts for the hierarchy of animal species: humans, by essence being spiritual, are at the top of the order, but if they do not reach moral perfection they are reincarnated in different creatures, in an order ranging from high towards low, namely in birds, land animals, reptiles, fish. This idea of a "scala naturae" has survived the millennia. It was not only thought that the species reflected eternal models, but that they show a gradual progress towards ever more complexity and spirituality.

Plato gives his version of how humans procreate and combines this with a remarkable theory about the heredity of intelligence. Procreation in the higher animals demands two sexes, and we can find here the pagan counterpart of the biblical text about the creation of a companion for Adam out of his rib: the soul of men who were during their lifetime cowards or who were unjust is born again in women.[9] Plato locates correctly the seat of our intellectual faculties in our heads, and then states that the marrow that comes out of the head is the semen. It descends along the spine and finally arrives in the sexual organs, where it causes sexual urge.[10] This implies of course that children fathered by men with good brains will be intelligent. In consequence this justifies that the most rational men, the philosophers, are the rulers and this by their birth (and education).

Pythagoras was still a monophysist: the Cosmos contains the Heavens, the Earth, the Gods, all living beings, plants, animals and men. It is one and it is the imitation of a mathematical order, though this underlying structure is hidden for the profane. Plato goes beyond this thought and proposes a dualistic conception: on the one hand there is the world of the phenomena, on the other hand the world of the ideas. The latter are the prototypes or models of the things we perceive and at the same time their cause. In a text from *The Republic* about the relation between the Sun and the supreme Good, he explains that the light of the Sun that is the child of the Good is the cause of the visible world and the source of the thriving of all things we see around us. Like the Sun stands above all things we can see and is their precondition, the idea of the Good Itself is also superior to all the other ideas of which it is the cause. As confirmed in the Timaeus, the Demiurge, who created the Cosmos, made it according to rational and mathematical principles and therefore it cannot but be the best possible world. We can all witness that in the concrete reality in which we live things are far from perfect, and this is especially true of human affairs be they moral, social or political. Yet, it is a consolation that apart from this corruptible, ephemeral world of individual phenomena exists a world of perfect, eternal ideas. Each of them is numerically one and at the same time it is the general form of all its imperfect material exemplifications, and those who are trained for it can see in them their ultimate source, namely the idea of the Good. In spite of this, Plato's portrayal of all things and especially of man is a sad one. Deficiency is inherent in this world, and it is the human condition never to match the ideal.

It is said that Plato was a malcontent man. Belonging to an old and noble family—he had an uncle who was one of the "thirty tyrants"—he was predestined for a political career, but seeing the corruption of the government and the discouraging state of affairs of the city he preferred to become a philosopher. Moreover, his beloved teacher Socrates had been brought to death under a democratic regime in Athens. He simply could not accept that the world he experienced was the ultimate reality. There had to be something else, a kind of justification for the shortcomings of the material world, a better world of which it was a mere reflection. In addition, this ideal reality founded the hope that things could revert to the happy state of a mythical Golden Age. Plato's philosophy and astral religiosity is wishful thinking par excellence, but also an admirable poetical and inspired creation of the mind. As Russell remarked in his *History of Western Philosophy*,[11] that compared to him, Aristotle had the mentality of a schoolteacher and shows himself as a man of poor imagination. Moreover, he completely lacked the religious fervour that gave wings to the thought of Plato.

Plato and the philosophers of nature

It is easy to understand that Plato hated the philosophers of nature, who doubted the existence of the Gods. He was aware of the danger the type of knowledge of which they were the proponents represented, for it was based upon sensorial experience and did not take into consideration a transcendent reality that could lend it a necessary structure and that could serve to explain why it is as it is. In *The Laws* he asks what can be the cause of atheism. It is in the first place the fault of the learned men that people reject the existence of the Gods:

> Now the effect of their compositions is this. When you and I produce our evidence of the existence of the gods, and allege this very point—the deity or divinity of the sun and moon, planets and earth—the converts of these sages will reply that they are but earth and stones, incapable of minding human conduct, however plausibly we have coated them over with a varnish of sugared eloquence.[12]

Before Plato's time, Anaxagoras, a philosopher of nature and a friend of Pericles, was condemned to death, when the statesman could no longer count on the favour of the Athenians and their rulers of the moment. The reason given for his condemnation was his atheism. He had indeed asserted that the sun and the moon are stones: the sun is a scourging rock, and the cold moon reflects its light. The reason for the fact a man could be condemned for such a statement is that, in contrast with Ionia, Athens had a centralised political regime with a state religion. As a consequence, patriotism and the cult of the Gods were considered to be inseparable. Plato defends religion as the very basis of society and names as its most dangerous enemy naturalism, more especially a naturalistic physics.

> ... Fire and water, earth and air—so they say—all owe their being to nature and chance, none of them to art; they in turn, are the agents, and the absolutely soulless agents, in the production of the bodies of the next rank, the earth, sun, moon, and stars. They drifted casually, each in virtue of their tendencies. As they came together in a certain fitting and convenient dispositions—hot with cold, dry with moist, soft with hard, and so on in all the inevitable casual combinations which arise from blending contraries—thus, and on this wise, they gave birth to the whole heavens and all their contents, and in due course, to all animals and plants, when once all the seasons of the year had been produced from those same causes—not, so they say, by the agency of mind, or any god, or art, but, as I tell you, by nature and chance....[13]

This is clearly a condemnation of the philosophers of nature, and amongst them especially the atomists. To them the order in nature had come about purely mechanically, atoms colliding with atoms and forming combinations of which the properties could be derived *quantitatively*. Given another initial state the world could have been composed of other entities and in this sense reality as we know it is the result of chance. However, not only these philosophers, but in his opinion also the sophists did incite to atheism. The most hideous scandal is the contention that the moral and political laws are neither ordained by the Gods, nor founded in a transcendent reality, but made by men and variable in different societies. It must be clear for every sensible mind that a spiritual reality must be the transcendent principle giving the values of society their ultimate foundation.

The soul

As we have seen, the reason for things to be as they are is that a Demiurge, a divine being, has created them out of the initial Chaos. The world is a living being with a soul. This world soul is the first cause and precedes and pervades everything. The Cosmos is a universal Civitas governed by the Good, which coincides with the Rational.[14] In this conception man, composed of an individual soul and a body provided with senses, is the link between two worlds, the spiritual world of the ideas he can reach for and the material world he lives in. Plato is at the origin of Western conception that the soul is more important than the body, and that the ultimate purpose of human existence is to leave the body behind and to go to heaven. Indeed, inspired by the mystical Pythagorean belief that the soul can free itself from the cycle of birth, death and reincarnation, he introduces a dual vision of man. His soul is like a bird in a cage; it belongs to the heavens and longs to return to its origin, but its body with its crude senses pulls it down. The metaphor is carried further; the soul, being birdlike, when enraptured by Eros, grows feathers preparing it to soar up. The tie with the body and its senses is what keeps it from its reunion with its divine principle.

> Did we not say some time ago that when the soul uses the instrumentality of the body for any inquiry, whether through sight or hearing or any other sense—using the body implies using the senses—it is drawn away by the body into the realm of the variable and loses its way and becomes confused and dizzy, as though it were fuddled, through contact with things of a similar nature?
>
> Certainly [answers Cebes, the interlocutor in this Dialogue].But when it investigates by itself, it passes into the realm of the pure and everlasting and immortal and changeless, and being of a kindred nature, when it is once independent and free from interference, consorts with it always and strays no longer, but remains, in that realm of the absolute, through contact with beings of a similar nature. And this condition of the soul we call wisdom.
>
> ... Now, Cebes, he said, see whether this is our conclusion from all that we have said. The soul is most like that which is divine, immortal, intelligible, uniform, indissoluble, and ever self-consistent and invariable, whereas the body is most like that which is human, mortal, multiform, un-

intelligible, dissoluble, and never self-consistent. Can we adduce any conflicting argument, my dear Cebes, that this is not so?

No, we cannot.

Very well then, in that case is it not natural for body to disintegrate rapidly, but for soul to be quite or very nearly indissoluble?

Certainly.[15]

It goes without saying that the nature of the soul is determining for the conception of man. The mystery cults, which promised a life after death had not been very successful, but always remained marginal. Also, in Plato's time the idea that the soul was eternal and could survive the body, if not novel, must have been attractive. The Elysium mentioned by Homer, which was situated on the banks of Oceanus, was also called the Island of the Blessed and reserved for favoured heroes. The Greeks generally believed that *if* there was anything after death for ordinary mortals, it was a shadowy form of existence somewhere deep down in the earth where Hades reigned, a place akin to the Hebrew *sheol*. Plato's promise of eternal life for the personal soul dwelling forever between the stars in exchange for an exemplary existence was more popular than the theories of the philosophers of nature in which death ended all.

In the text cited above, we are told by Plato that though human beings certainly *inhabit* a body, they do not *coincide* with it. The soul is the master made to command, while the body is the rebellious slave that must be forced to obey. In fact, Plato elaborated fundamental ideas already present in Pythagoreanism. The God of his philosophy is the Sun, Apollo; the mystical experience that can free a man from his bodily constraints is ecstasy, which means literally that he reaches a state where he is outside himself. This experience is sharply to be contrasted with the Dionysian vision, where man, through intoxication, knows a rapture that strips him of the awareness of his self, enabling him to lose the consciousness of his individuality and to make him sense his union with nature.

Plato has shaped the Western conception of the relation between body and soul. The reason why Plato denied the identity of a person with his body is complex.[16] He took over from Socrates the thesis that a person always wants to act for the best. This was to ask for trouble, as this is not at all self-evident. Also, he had to find good arguments in order to be able to reject the many examples that counter this thesis. A great part of what we do seems to be neither bad nor good. It could be said, however, that for example a man quenching his thirst is perhaps not doing something that could be called morally good, but from Plato's teleological viewpoint it can be argued that his behaviour is in accordance with the general good that is at the base of the total reality, as in this way he sustains his life. More difficult to explain as behaviour for the good is intemperance: to drink when one is thirsty is rational, but somebody who gets drunk cannot be considered to act rationally. What about consciously committed brutal and cruel acts, or what about lust and greed? That the cause for them is always ignorance seems far-fetched: more commonsensical is that in many cases people do bad things simply because they like to do them, or because they hope to gain something by acting that way. It could be supposed that when Plato said that we all act in order to achieve the good, he meant that, when having to make a decision, as human beings whose essence it is to be rational, we consciously compare alternatives and make choices. Following this line of reasoning, a solution could be to suppose that if we make a wrong choice we were enticed to it by our body, and that, overcome by its emotions or desire, our soul was defeated. In that case we did not act knowingly in the sense of taking a rational decision, which is the same as taking it willingly. In other words, we did not behave as human beings, but like animals obeying the urges of their bodies.[17] For this interpretation to make sense it must be presupposed that we have a dual nature and are composed of two substances and not one. At first sight this was not a problem, as the soul was eternal, the body mortal and this implied they were of an utterly different nature. To separate them explicitly in this way was a decisive step to take, which caused immediately a problem: if body and soul are two substances that are only temporarily linked to each other, what exactly is their relation?

The problem remains whether the seat of emotions and desires is not the soul rather than the body. Plato has proposed a tripartite division of the soul[18]; it has different desires, namely the desire of reason, the desire linked to passion and the desire related to physical urge. The image he uses for this threefold composition of the soul is that of a charioteer and his team of two horses, one horse being noble beautiful and complying, the other wild, ugly and unruly. Passion gives the soul wings and makes it soar upwards, vile lust drags it down. The intellect brings measure, the good life being that of a man who has found an equilibrium between the opposite tendencies. Of course, it is difficult to bring this vision in accordance with the argument that when we do wrong it is the body and not the lower part of the soul that is to blame taking bad decisions. Though everything seems to be well-reasoned in Plato's philosophy, perhaps we must allow for a number of inconsistencies.

Aristotle

His rejection of dualism

Aristotle was both a philosopher and a scientist, which cannot be said of Plato. It must be conceded that the latter knew the mathematics of his time, but he was not creative in this domain. Aristotle had the temperament of a philosopher of nature, but he became the pupil of a religious philosopher for whom the salvation of the soul was a central matter. He had to reconcile the teaching of Plato with his own interests and conceptions, and from these conflicting elements a new type of philosophy arose. As was to be expected, the influence of Plato was very strong in Aristotle's younger years, but it gradually faded as he became more self-assured. His growing criticism was mainly directed against Plato's dualism. In his *Metaphysics*, he addressed the problem of the existence of the ideal forms, and he came to the conclusion that their realm was totally overcrowded. He cut away all the ideas he thought to be redundant, ideas for qualities, for relations, for negations, for dyads, till only ideas for substances were left. In fact he declared that reality was exclusively composed of first substances, things that could exist autonomously, i.e. concrete individual things. The qualities of these things were secondary substances, which existed only to the extent they are inherent in the first substances. the latter are moulded upon an eternal form which is reflected in the fact that things of one kind show the same essential properties. The sums of these properties are not a separate reality, but exist only as intrinsic parts of concrete individuals. Aristotle reconciled in these individuals matter and form of which neither can exist on its own. He found a similar solution for the dualism of mind and body; they form a unity that is only destroyed at the moment of our death.

The nature of the soul.

The belief that the soul is an eternal substance inhabiting temporarily a body and having the task to keep it in check stems from Plato. He is the source for the tripartite model of the soul that has survived under a different and far more elaborated form in Aristotle's theory on the subject. In earlier writings Aristotle seems to have still more or less adhered to Plato's separation of body and soul, but in *De Anima* he is very clear: the individual soul and the body are one. Therefore, when he proposes a theory of the soul consisting of three parts, we must not think of a physical division of the brain into three cells or compartments. The reason is twofold. The first one is that, while Plato correctly thought that the seat of the soul is in the brain, because the highest place is most suitable for it as its temporary dwelling, Aristotle, though he knew a lot more of anatomy, wrongly supposed that the seat of the soul was the heart and that is clearly not divided in this way. The second reason why we must not think of a physical division is that the soul is in the opinion of Aristotle the completely immaterial complement of the body. Living beings, like all other substances, are composed of matter and form. A body that is potentially alive is like a machine ready to function; it is the soul that turns it into an actual living being and in this sense makes out its finality. Also, the soul takes part in all that happens in the body and to the body. However, a problem remained. Aristotle honoured the accepted principle that interactions suppose either that the elements show a certain likeness, or else that there must be a medium. We have here two realities; one, the body, is extended, the other, the soul, is not. It seems he had no good explanation of how exactly these different realities combined with each other, but he simply posited their unity as counterparts of each other. It could be that he felt uneasy about this and this could be the reason why he splits off from the soul a part that is not individual but general and free from material ties. From the unity of body and soul follows that he could not but reject as absurd the belief in the transmigration of the souls.

The Aristotelian theory of the soul is complicated and I shall follow in broad lines O. Hamelin, who has given a particularly clear exposition of it.[19] According to Aristotle, the function of the brain is to cool the blood, and the soul is not to be located in the head, but in the heart, which can be easily understood, the latter organ being the "animator" of the animal. Further reasons for choosing the heart as the seat of the soul are the following: the heart contains the "pneuma" or breath that is the source of animal warmth, and when it ceases to beat the animal dies. The system of veins and arteries has as its centre the heart, and, as animals can bleed to death, blood seems to be the life giving substance by excellence. All sensations go from the sense organs to the heart from which all movements are initiated.

The soul has broadly speaking two functions, to provide us with thought and to initiate movement. Where the first of its functions, *thought*, is concerned, as mentioned earlier, it can be divided into three parts. It is important to note that, in contrast with Plato, Aristotle did not construct his theory in terms of opposites, but in terms of hierarchical order.[20] The first and lowest part is the *vegetative soul* that is responsible for the working of the body, for its nourishment, and for survival and reproduction. It is present in all living beings, plants, animals and men alike, and it works autonomously. Indeed, without conscious incentive it can keep all the vital functions going on in plants and in animals, and even when the latter are asleep their heart pumps and they continue to breathe. To possess a *sensitive soul* is what defines an animal, as indeed even the lowest animals at least are sensitive to touch. Aristotle adds to the different senses the "*sensus communis*" that is their substratum. Being indeterminate itself, it lends the necessary unity to the specialised senses. Moreover, it enables animals to be conscious of their perceptions. The *intellective* soul belongs only to man and enables him to think abstractly. For understanding this rational soul we must start with memory, because it is the precondition for intelligence. (Memory is not dealt with in *De Anima*, but elsewhere). Its precondition is imagination, as in order to recall, animals must be able to represent mentally what they have perceived.[21] Memory is a basic capacity of the soul, in hierarchical order immediately superior to the senses and just under the intellect. It consists of traces left by perceptions in the sensitive soul or rather it is caused by the continuation of movements triggered by perceptions.[22]

Let us now enter into further detail where the intellective soul is concerned. The soul has two main functions: thought and the production of local movement. We begin with thought. In fact the vegetative, sensitive and intellective soul are not different parts, but form a continuum, the lowest part can exist without the other two, but the second part not without the first, the third not without the two lower parts. Though most animals are capable of representing things for themselves and remembering them, a special form of memory, though not part of human intelligence in the strict sense, is a capacity acquired only by human beings: we can imagine at will all kind of things without relation to the actual situation we are in. However remarkable, this type of thought does not yet attain the first level of what the intellective soul can achieve, namely judgement that is universal and implies truth or falsity. Aristotle rejects Plato's theory of reminiscence of what the soul once saw when it was in the realm of the stars between two reincarnations. All knowledge must on the contrary start from experience, i.e. from perceptions of individuals, though such perceptions are not yet knowledge, because they lack generality. Following Aristotle, knowledge presupposes already knowledge, but in the case of the lowest level, where we must start from the concrete and individual, it can be only potentially present. In other words, perception of particulars is not real knowledge, but the experience of them contains in potency comprehension of the universal.

The intellect is either *passive* or *active*. The passive intellect is like a tablet on which originally nothing was written, but which is a receptacle for sensorial inscriptions. This guarantees the possibility of objective knowledge, its source being outside the subject. The passive intellect must be known by the active intellect and must be actualised by intelligible forms or general concepts. This is in accordance with Aristotle's thesis that soul and body are one; as human cognition presupposes sensations and representations or images, the passive intellect is always linked to the body that provides them. The mind is passive to the extent that "it becomes all things," namely all that enters it, all its objects. On the other hand it is active, because "it makes all things," like light that makes potential into actual colours. This is straightforward enough, but in a crucial passage of *De Anima*, he adds:

> Mind in this sense is separable, impassive and unmixed, since it is essentially an activity; for the agent is always superior to the patient, and the originating cause to the matter. Actual knowledge is identical with its object. Potential is prior in time to actual knowledge in the individual, but in general it is not prior in time. Mind does not think intermittently. When isolated it is its true self, and this alone is immortal and everlasting (we do not remember because, while mind in this sense cannot be acted upon, mind in the passive sense is perishable), and without this nothing thinks.[23]

Indeed, we do not remember the activities of the mind in the past in between different reincarnations like Plato had suggested, which presupposes that the soul is eternal. On the contrary, it is tied to a material body that lives a life finite in time. However, Aristotle suggests here that the active intellect not in general, *but as such*, in its pure aspect, is immaterial; it is pure form and thinks without needing an organ for it.[24] How the active mind *as such* can be distinguished from the active mind in general remains unclear. Anyway, the idea of a soul that is not tied to the body and thereby not individual but general and that could be immortal has left room, as we shall see, for speculation and diverse interpretations, one of which is that it is to be identified with the Divine Mind. Nevertheless, it was very difficult for orthodox philosophers to accept the Aristotelian vision, because for the Christian if someone is redeemed by the grace of God, it is his individual soul, given to him by his Creator, that will live further on after his body is decomposed. Though the pure intellect of Aristotle was immortal, it was not a satisfactory solution, because it was a kind of abstract spirituality that had nothing to do with the personal soul God breathes into each of us. The discussions about the mortality of the soul in the thirteenth century will be continued in the Renaissance and thereafter.

We must also consider the other function of the soul, that of *motor of our actions*. When we act, in many cases emotions are involved. The only affective phenomena taken into account by Aristotle are pleasure and pain. It is remarkable that the emotions are not a distinct part of the soul. Perhaps this can be explained by the fact they are only derived in his theory. Pleasure reveals the presence of appetence, which is anterior to it and its foundation. What we want is the accomplishment of a function giving us pleasure and not in the first place pleasure itself. The latter depends upon the natural activity it accompanies and upon its modalities, and one could say that pleasure is an additional effect of the act. The practical intellect makes use of desire, and in a sense this is the matter of which this intellect is the form. Practical intellect cannot be absorbed by the primary cognitive part of the soul, and it remains to a certain degree irrational. Desire and avoidance are compared to affirmation and negation. We want something, because we represent it to ourselves as desirable, we avoid something because we represent it as dangerous, painful, disgusting, etc. Aristotle's approach is rationalistic: the supreme intelligible is the supreme desirable and something we want is good if it corresponds to the rationality and harmony of the total reality. The example given by Hamelin is the following: it is necessary that we drink; this is drinkable, and the animal drinks it. However, the desire can be induced by a false representation of what is desirable, and the source of this can be a mere opinion instead of the practical intellect. In that case the desire is not rational, but irrational and it can command us, taking the place of the practical intellect, like in the case of intemperance. It is in this discrepancy between the rational and the irrational that contingency arises. Aristotle believes in free will, but it is not really clear where lies the distinction between freedom and a mere chance event. If we compare his conception with that of Plato, who declares that we always act for the best, we can see that Aristotle, as in other matters, was more realistic. He was aware that, though in principle individuals act following the finality of their being that is part of the finality of reality as a whole, they do not always realise completely the perfection of their nature.[25]

The tripartite division of the soul proposed by Plato has been adopted by the scholars of the Middle Ages and this division has been projected upon the brain. This is a departure from the original concept of Aristotle, who believed the seat of the soul was the heart. In C. Blakemore's *Mechanics of the Mind*, we can find illustrations from the Middle Ages and the Renaissance showing different representations of the brain composed of three (or sometimes more) cells. One of these images is a kind of hybrid model, with as the seat of our psyche a combination of the brain and the heart. The brain is connected with the sensorial organs, the heart with the brain and curiously with the ears, but not with the other sense organs.[26] Though today nobody believes we think with our heart, the idea that it is the centre of our emotional life is common to most of us. We still speak of a person that is good as having "a good heart," we say that "our heart is broken" and that we do something "with pain in the heart," we swear an oath with our hand upon our heart, etc. Probably the reason is that, though we do feel emotions *with* our brain, we do not feel them *in* it, but in the heart region.

Aristotle's naturalistic conception of man

We can classify Aristotle's work under four headings: physical, logical and metaphysical, ethical and political, and biological. According to Benjamin Farrington, in his *Greek Science*, his greatest merit as a scientist must be sought in his biology.[27] Probably this is due to his new attitude to nature and to observation. Though the teleological view he inherited from Plato is always present in his work, in his biology he explicitly states that the knowledge of living beings in spite of Plato's objections should be based upon observations. Aristotle defines man as a rational animal, but still he is an animal. Though he considers that science and philosophy are the most worthy occupations for a man, especially if he is of good birth and can afford to enjoy plenty of leisure, neither philosophical studies, nor living a moral life guarantee life after death. The function of the soul is first and foremost to think and to command action and its purification is not a central concern. Man has become an object of observation, rather than of speculation.

In his works on biology, but mainly in his *Historia Animalium*, we can find a detailed description of the special physical and psychical properties of man. Aristotle has not made a proper classification of the animal species or rather he has offered us several possible classifications. According to specialists it was never his intention to construct a definitive taxonomy, but to give a factual survey in which the differences between the species are indicated. This should enable him to reach his ultimate aim that was to find the causes (material, formal, efficient and final) for the empirical facts. Nevertheless, he considers several possible criteria that could be used for classification, and he compares their merit. For example, he comes to the conclusion that ordering by one characteristic comparing different types of locomotion is not efficient, because this criterion cuts across the species and the result is a great deal of overlapping.[28] A better criterion would be the degree of perfection of the offspring at the moment it is born, which implies several characteristics. As I have explained,[29] Aristotle rejected the method of division by opposition and started from essential properties to catalogue elements of reality in general. To start from opposites would divide natural groups; for example water-animals versus land-animals would divide the group of birds.[30] In consequence, in his tentative orderings of the species he does not obtain a chain of ever more perfect beings, like Plato, but rather a treelike structure with bushes of groups splitting up into subgroups and of subgroups splitting up into their species.

What is of interest to us is that man, when considered biologically and not as a rational being, is not a kind apart from the other animals. Thus, according to the criterion of the degree of perfection of the offspring, our species is put together in one group with the ox, the horse, the dog and all the hairy animals, "those quadrupeds that put their hind legs inward" and the Cetacea, in other words with the mammals.[31]

All the species are separately created by Nature that uses matter for this process and lends it its form. They are the result of a judicious arrangement of different proportions of matter that is warm, cold, fluid or solid. It is heat that plays the predominant role and if present in the right proportions with fluidity this gives the best results. Describing the animal species, of which he mentions several hundreds, he accords to them different degrees of perfection. The problem is that the teleological principle he accepts implies that our world should be the best possible one. How then can it be that Nature sometimes creates deficient beings or even monsters? This can be explained by the fact that Nature has to work with matter that in itself shows no perfection and that it is not always possible to impart it with the movements in such a way that matter is kept fully under control. Though Aristotle has not given an answer to the question that readily comes to the mind why in a world that is a rational order does imperfection exist, it is probable that he would have accounted for the less perfect species by stating that all the species must in one way or another have some usefulness. The best examples of an unavoidable element of imperfection are the females that are the result of the straying of Nature, but are nevertheless necessary for reproduction. As he puts it:

> The first beginning of this "straying" is seen when a female offspring is produced instead of a male. If the seminal residue of the female parent is well concocted (i.e., has been properly acted upon by "the hot"), then the "movements" supplied to it by the male parent will fashion it after the male's own shape, i.e., will produce a male offspring.[32]

It goes without saying that man is the most perfect being. His rationality sets him apart from the rest and there is no intermediate form of life between him and the animals. As an animal, however, he is only one of the intelligent species. Aristotle explicitly states that nature proceeds by very small steps and that there is a great continuity between the species.[33] The diversity is not due, however, to evolution, but to Nature creating *separately* a wide range of living beings. The first to appear are plants that seem animate compared to inanimate matter, but inanimate compared to animals. The change from them to the simplest animals is continuous, and gradually the animals show more and more life and movement. Plants reproducing themselves by seed have no other apparent function than to make other ones like themselves and there are animals that similarly seem to have no other function than generation. However, when in animals sensation is added, their life changes; they mate for the pleasure that accompanies it and they will rear their young in other ways than the senseless animals. Some nourish their young and care for them, but afterwards they leave them on their own, others have more understanding and also memory and continue the association and social relations with them. The energy of all animals is spent in reproduction and in the search for food. And Aristotle concludes: "what is natural is pleasant, and all pursue their natural pleasure."[34]

The continuity between the species explains that Aristotle has no problem to admit that other animal species have types of intelligence that come near to that of man or are analogous to it.[35] What is uniquely human however is that he is deliberative, as, indeed, many animals can learn and remember, but he can remember at will. Most important is that he can speak. He is also the only animal able to laugh and cry. He laughs when he is tickled under his arms. His psychical capacities are sometimes related by Aristotle to anatomical characteristics. Thus his intelligence goes together with the fact he has hands. It is well-known that Empedocles believed we are intelligent because we can use our hands. To Aristotle it is the other way round: in his teleological view, because we are intelligent we must be able to perform certain acts for which hands are required. He provides us with an extensive description of the human physiognomy. For example, the parts of the head, seat of the senses and of our speech organ, are meticulously described. Man is the only animal with such a thick coat of hair on the head and this because his brain (the organ cooling the blood) has to be protected from heat. Of all animals he has the most variable eye colour. We could continue with the examination of the mouth, the teeth, the tongue, the nose, the ears… The other parts of the body are also treated in detail and compared to similar parts in animals.

Naturally, most interesting for us are the differences he mentions that are relevant from a modern point of view. Aristotle gives us a most accurate analysis of different species and he must have had the occasion to observe a number of specimen of them. One of the special characteristics of man is that he walks erect, and further that when he walks he looks straightforward, which is indeed only possible because of the position of the head relative to the spine. Linked to erect locomotion is the fact the lower limbs in humans are much more developed, heavier and stronger than in other species. To walk erect is a precondition of intelligence. He is aware of the fact that women have much more difficulty to give birth than other female animals, but as far as I know he does not connect this to the form of the pelvis and also to the big head of our babies. All these distinctive characteristics of man are to be found dispersed over the *Historia Animalium*, and do not as such form a subject apart. However, everything becomes clearer when Aristotle compares apes with men. These are his conclusions. After having drawn the attention to the differences between apes and monkeys, which both have a dual nature, sharing characteristics of quadrupeds and man, he says the following:

> Apes are hairy on the back, in virtue of their being quadrupeds, and hairy on their fronts, in virtue of being man-like (the distribution of hair is a point in which the arrangement in man is the reverse of that in the quadrupeds), except that the hair is coarse; hence apes are covered with thick hair both back and front. Their face shows many resemblances to that of man: it has similar nostrils and ears, and teeth, both front teeth and molars, like those of man. Furthermore, whereas the other quadrupeds have no lashes on one of the two eyelids, like men the ape has them on both, though they are extremely fine, especially the lower ones, and very small indeed. Also, the ape has on its chest two teats, though the breasts on which they are placed are small. It has arms like a man, though covered in hair, and it bends its legs as a man does, i.e., with the convexities of both limbs towards each other. Besides this, the ape has hands, fingers and nails like a man, except that all these parts tend to be less refined. Its feet, however, are a peculiarity. They are like large hands and the toes are like fingers, the middle one being the longest; and the lower part of the foot is like a hand, and extends towards the extremities, like a palm; and this palm towards the end is harder, in a poor and indistinct way resembling a heel. The ape uses its feet both as feet and hands, and clenches them like fists. The upper arm and thigh are short compared with the lower arm and the calf. It has no protruding navel, but the part where the navel should be is hard. Its upper parts are much larger than the lower; as with quadrupeds: the proportion is about 5 to 3. Because of this, and because it has feet which are similar to hands and consist as it were of hands as well as feet…, because of all this the ape spends most of its time on all fours rather than upright. The generative parts of the female are rather similar to those of a woman; those of the male are more like a dog's than a man's are…. In all animals of this sort the internal parts, when dissected, resemble those of man.[36]

Certainly there would be much more to say about this subject: Aristotle was the first to offer us a very accurate and complete biological description of our species. It must be understood that this description is an integral part of what we could consider as an encyclopaedia of the animal kingdom. The contrast of this naturalistic approach with Plato's speculations could not be greater.

Neo-Platonism. A long-lasting world view

Before closing the chapter about antiquity we must consecrate a section to Neo-Platonism, because it is one of the constituent elements of the dominant idealist tenet in Western thought. The most powerful Neo-Platonist vision was that of Plotinus. He was not only a follower of Plato but also a mystic, and this was a magical combination. He created a view of God and the world, which, though pagan, proved to be highly suitable for a Christian interpretation and which was an enduring source of inspiration for idealist philosophy. Plotinus combated different Christian sects, more especially Gnosticism. He did not want to create a new philosophy, but to understand Plato, using often the commentaries of Aristotle, as he had known the master and knew his work from firsthand. His influence on the theology and philosophy of the Middle Ages was most important. Neo-Platonism was a source of inspiration for the Islam and for Jewish religion. In the Renaissance Platonism knew a revival and his influence on Marsilio Ficino and on Pico della Mirandola is well-known. We can recognise many of his ideas in German idealism, especially in the work of Hegel, but also in that of other idealists, such as Henri Bergson and Teilhard the Chardin and the philosophy of Heidegger.

In the first century A.D., there had been a revival of Pythagoreanism. The Pythagorean theory of the soul met the emotional need to have certainty about what happened after death. Ascetic morality, to repent and abstain from sensual pleasure, was linked to the promise of eternal life. The destiny of the soul was to return after a cycle of reincarnations to heaven where it originally came from and this was in fact a promise of an eternal stay amongst the stars. Next to this emotional need there was also the intellectual demand to understand the nature of transcendent reality, and according to Neo-Pythagoreanism the answer had to be found in a fantastic numerology that in an abstract way indicated which were the first principles behind the world we can perceive.

The diverse forms of reality are the different degrees of realisation of the original "One" that transcends being and essence. The first degree under the One is Intelligible Being or the Ideas, the second is the Soul participating in the Ideas and the third is Matter. (Whether this was really the original Pythagorean world view is matter of debate; it could be based on a commentary by Simplicius on Plato's dialogue *Parmenides.*) Anyway, this vision of the world and the theory of the destiny of the soul became the core of Neo-Platonism.

We must keep in mind that it was not so much the intention of the Neo-Platonists to use this view of reality for explaining the perceptual world as to escape from it in a transcendent reality that was eternal and not liable to corruption. Only its contemplation could make a person happy, while the senses could bring nothing but misery. Plotinus, who lived in the third century A.D., was the most renowned of the Neo-Platonists. His disciple Porphyry published after his death the orally communicated theories of his master on different subjects, under the name of *Enneads*; i.a. the usefulness of astrology, how the soul descends in a body, is united to it, the problem of the memory in various kinds of souls, from the human soul to the World Soul. These subjects must be seen in the light of the coping stone of all this, a world vision dominated by the triple unity of God, Destiny and Nature.[37]

According to Plotinus, the principle of the hierarchical order in the world is that the perfection of a being depends from the degree of unification of its parts. For example, a heap of stones has only a feeble degree of unity and a choir or an army forms a loose collective body of individuals, but a stronger coherence can be found among the parts of a living being, because they are kept together by the tension of its soul. Yet, the parts of a living creature are juxtaposed and occupy different places, but in science a specific theorem, being one of its parts, contains potentially all the other theorems and there is no such differentiation in space. This makes the unity of a science superior to that of a living being.

We can see from this how Plotinus passes from dead matter to living matter and from the latter to the spiritual realm, always using the principle of unity. We cannot understand his philosophy properly without keeping in mind that his reasoning is dialectical, rather than demonstrative. All which is not perfectly one presupposes a reality superior to it of which the parts are more united; thus the parts of a living body or of the world presuppose a soul and the World Soul respectively, and the theorems of a science presuppose a divine intellect capable of grasping it. The ultimate reality keeping everything together is the One that is the principle of Being. The unity of the One is not a formal emptiness, but contains in potency all that will develop from it.[38]

This unity of the various elements of reality is never purely concrete, but it is of a spiritual nature; it is actually the contemplation of what is a reality above it that lends the parts of a thing their coherence. Thus, the soul of a living being contains an inseparable seminal rationality of the parts, the plan and the details of the body. Being is always subordinated to the One of which the true nature is contemplation. Even Nature is speculation; it is silent, mute, unconscious contemplation of the eternal models reflected in it. Each superior principle remains immobile in its incorruptible perfection; nothing of it leaves it to enter what is inferior. Therefore, it is not a real agent, but can be compared to light that illuminates and is reflected by what it shines upon, but only to the extent that this object is capable of being illuminated and of reflecting light. This is a crucial idea in idealism. Thus Hegel says in his *Wissenschaft der Logik*[39] that the proposition that what is finite is not material, but ideal, is the core of Idealism. Its correlate is to deny that the finite and concrete is real being. When Ionian philosophers see matter as active, as that what constitutes and produces the world order, they consider it as containing a first principle, and therefore, fundamentally it is not concrete, but abstract and thereby ideal.

The Cosmos is one, finite and eternal; it is unique, unified by the sympathy of its parts and with as its centre the earth. To grasp this is the precondition to understand the three hypostases or principles of being. The first hypostasis is the One that is absolutely indivisible, and of which like the One of Parmenides all can be said and all can be denied, because it is above Being and Essence. The One, being neither essence nor substance and having itself no positive properties, is also the supreme Good. It gives each thing existence and it is at the same time the ultimate aim of all that exists. However, this is true only in a manner of speaking, because it is nothing, not even one, nor good, but only a supra-essential non-being. The great question is "why did it not remain unique, contracted in itself?" Or what is the same question, "why is there Being rather than Nothing?" The answer of Plotinus is that all that is perfect is productive, and he compares this to a living being that reaching its completeness in the adult stage must produce a similar being. This production of Being by the superior principle is unconscious, involuntary, due to an excess of richness. It has been likened to the overflow of a source and has been called *emanation*, or, as Plotinus himself termed it, *procession*. The product wants to stay as close as possible to what has produced it and turns back to contemplate it, which is *conversion*. Martin Heidegger in his *Was ist Metaphysik?*[40] asked the same question, and like Plotinus, sees himself as going back to the pre-Socratic philosophers. His criticism of Western metaphysics is that it is about the general characteristics of beings, instead of about Being. Being is to him, what the One is to Plotinus: it is the first ground, *Urgrund*, of everything. In ordinary life we are not aware of it. We cannot grasp it, let alone know it. Nevertheless we long for it, though it is a transcendent reality we cannot reach it–he compares it in this respect to the horizon that beckons us; we can go towards it, but it remains always unattainable–but it reveals itself in exceptional existential experiences. Descartes compared science to a tree with metaphysics as its roots. Instead of looking for them, we should be looking for the soil in which the tree grows, we should be looking for the *Urgrund.* Neither Plotinus, nor Heidegger called their source of reality explicitly God. Teilhard de Chardin, however, who was a theologian, introduced his idea of le *Point Oméga*, or the source of all reality, as being identical with God. Like Plotinus, he believed humans are capable of turning back towards the source of all being, thereby closing the circle between alpha and omega.

The second hypostasis is Being, the Intelligible World and Intelligence, a triad forming a complete world in itself and not to be understood as an abstract scheme of the sensible reality. Reality, indistinct in the One, spreads in all directions in a hierarchical multiplicity of genera and species. Plotinus, unlike Aristotle did not conceive of this multiplicity according to a logic based on essences, who added to genera specific differences for defining the species, but like Plato, according to a dialectical logic proceeding by division. The difference on the ontological level is that to Aristotle the species were simply sums of separate individuals and the genera sums of species, while Plato sees in the genus, a whole, a separate reality that is divided into species and he sees in the progress from the genus to the species not an enrichment, but a transition from the whole to its parts in which the fullness of the former is retained. For Aristotle the individual is the concrete realisation of the specific form, a *species speciallissima*, with its essential properties, plus an undetermined number of other characteristics, and to him the individual is more real than the species. Socrates is more than the concept "man"; he is *this* man, and in contrast to the concept he cannot be defined. The empirical reality is here more real than the intelligible reality, which exists only inherent in it as a sum of essential qualities. Plotinus is faithful to Plato and claims that the intelligible has primacy over its concrete realisations. He denies that the form, in order to be concretely realised in individuals, has to be augmented with positive characteristics extraneous, not essential, to it. Nevertheless, such characteristics undeniably are present, but to solve this problem Plotinus states that they are also expressed in the intelligible world, because it has to be as rich as possible. Even sensations in the material living being retain something that is spiritual and stems from its intelligible origin. The sense organs are neither the result of a chance process nor of a providential plan, but they are an imperfect imitation of a higher reality.

For completeness sake it must be said that the "Intelligibles" or Ideas are *in* Intelligence and not outside it. The reason is that the reality that is above Intelligence is not the world of ideas, but the One, and as we have seen the One is not intelligible. It is therefore *in itself* that Intelligence must find its complete richness, neither in an inferior reality, nor in a superior reality. Thought of itself gives the certainty of its own existence, and an immediate knowledge of its content. It is here that we find the source of the *cogito ergo sum* of Saint Augustine and of Descartes.[41] Intelligence is contemplation and vision of the One and it is at once knowledge of itself, and what is the same, of the intelligible world.

The One produces, next to Being and Intelligibles, Intelligence, and Intelligence produces the third hypostasis, the Soul. The worldview of Plotinus is not in the least mechanical, on the contrary, and at the difference with Aristotle, the movements of bodies in general are not governed by their essence, stones falling to the earth because it is their nature to do so, men walking on two feet for an identical reason, and celestial bodies moving in circles, because they are divine. According to Plotinus it is their soul that moves them. The World Soul reigns over the general destiny of the world, and the individual souls, the egos of men, which are intermediate between the world of the ideas and the sensory world, determine what happens in ordinary life. As the explanatory principle is the unity of the world, the individual souls of men and the souls in the stars must be understood as fragments of one totality, the World Soul, which has a cosmic role. The third hypostasis is the mediator between the intelligible world and the material world. Proceeding from the spiritual reality, it is turned towards the higher reality it comes from and that it contemplates, but it also touches the material world, because it orders and governs matter by exerting an influence that emanates from it. While it remains purely contemplative, nevertheless it acts.

However, we must not ascribe to it an intention to realise the Good, or to save mankind.

The only real Gods were according to Plotinus the eternal souls ruling the stars. He accorded an extreme importance to incantations and prayers, to astrological predictions and to the cult of statues and believed in their efficacy if the traditional procedures were respected. However, the incantations were not directed to a God, but to the metaphysical principle of sympathy relating all the parts of the world and linking the individual person to the totality.

The last and fourth hypostasis is Matter. It must not be understood by its relation to form, but in itself. It is indeterminate and indeterminable. Form does not give it properties, and when form leaves it, it remains as indeterminate as before. To the ancient philosophers of nature matter had been an active principle, but not to Plotinus; for him it is absolute poverty. There is no real unity between matter and form, like Aristotle had taught, but the sensible concrete reality is only the reflection of form in matter. Matter is destitution, want, incapacity, of which nothing positive can be said. This is absolute evil; it is not an imperfection only in comparison to what is superior to it, but evil in se. From the moment the soul relates to matter it partakes in change and corruption. The weakness of the soul, its impurity, all that is bad in it, is due to its contact with matter, and we must not try to dominate it, but to flee from it altogether. If matter nevertheless exists it is because all the degrees of reality must be exhausted. It is not, as could be thought, independent from the One, but "like a last reflection, before the complete obscurity of Non-Being."[42] In his last writings, Plotinus develops the idea that evil does not really exist, because rationality is the ordering principle of the world of which evil is an integral part and therefore there must be a justification even for the lowest degree of reality. Each creature has its place in the vast harmony and whatever its destiny, whatever its incapacities, its sorrow and suffering, from the point of view of the total reality its existence cannot be absolutely bad for the universe. On the one hand asceticism was preached, and we must turn away from the worldly matters that can harm our soul, on the other hand stoic acceptation of our destiny was demanded.

This world view contains a vision of man that has never again completely disappeared from theology and idealistic philosophy. Though it was a cerebral approach and in fact completely alien to the Biblical conception of man, it has survived in the Western idealist tradition. Its two main propagators have been Saint Augustine and Pseudo-Dionysius, who have transmitted it to the Middle Ages. Scotus Eriugena, Bernardus Silvestris and after them many others have been inspired by this philosophy. Robert Grossesteste's light-metaphysics is modelled upon it. In the cosmology of the latter the One, or God, has become pure light, concentrated, drawn together into itself. The emanation is radiation of light in all directions, creating time and space in the process. Light is the carrier of the forms of the species; what was one, becomes diverse when it concretely realises itself; light travels along straight lines till it reaches the limits of the universe, where it is reflected inward, and returns to its origin, the One. This beautiful vision is reworked by his pupil, Roger Bacon, who like his master, believed that mathematics (optics, perspectivism), is the key to the knowledge of the world. The tradition is like a strong undercurrent coming to the surface time and again.

The central question is what becomes of us after death. Plotinus explains that all the individual souls are part of the World Soul, and that for each of them is prepared a place that is in accordance with its nature. It has to rule over this place during the time of its stay, which is determined by the general order of things. The soul, as we have seen, has two main aspects: it is turned towards the intelligible order and is thereby intelligence and stays with intelligence, but a reflection of it illuminates and vivifies the body. If the soul turns towards this reflection, and like Narcissus, attracted by his own image, drowns by trying to embrace it, the soul precipitates itself towards it, and from then on it shares with the body whatever it experiences. This is the *descent* of the soul, or rather of that part of it, called, "me" or "self." Also, its ultimate destiny will be in accordance with the sin it has thus committed. In this way in the fixed world of Plotinus is introduced a historical dimension, related to the different ontological levels the soul can turn to.

In the world of the senses it can lead different kinds of life. In hierarchical order: a life of pleasure, of activity, of reflection where it becomes master of itself, or of intuitive thought where it becomes intelligence. The intellect can turn itself towards the Intelligible, but the One cannot be rationally understood, but only be revealed in a vision that goes beyond the intellectual capacities. The highest level the soul can reach for is the One, but this can be achieved only in a mystical rapture, an ineffable experience in which the soul is no longer conscious of itself and has a direct contact with the first principle.[43]

In this world vision, man is essentially a soul, capable of the contemplation of the intelligible world, and forever longing to return to its ultimate source, the unifying principle, the One. However, the nature of his soul is such that as an ego he can also long for its material reflection, descend in a body and experience the material world. By these two aspects of his nature, man not only mediates between the spiritual and the material, but is the mediator between the alpha and the omega, the beginning and the end. The theory of emanation can easily lead to pantheism, but from the point of view of the orthodox Christian doctrine in which God does not coincide with his creation, but is transcendent, this is an aberration. At most pan*en*theism is reconcilable with the Christian Neo-Platonist conception of the world, because it does not identify God with the concrete reality emanating from him, but only says that God is present everywhere in the world.

The extraordinary attraction Neo-Platonism has exerted must probably be explained by the fact that, if adjusted in the right way, this heathen world vision was as it were ready for use by Christian philosophers and theologians as the intellectual backbone of faith.

Notes

1. B. Farrington, *Greek Science. Its Meaning for Us*, Penguin Books, Harmondsworth, 1944, 29.
2. Cf. J. Tricot, *Aristote. La Métaphysique*, Vrin, Paris, 1970, 58–59.
3. *Ibidem*, 58.
4. Cf. F. Nietzsche, *Die Geburt der Tragödie aus dem Geiste der Musik*, in: *Nietzsche Werke*, K. Schlechta, ed., Hanser Verlag, München, 1956, Vol. 1, 21 and following.
5. B. Farrington, *Greek Science*, 42.
6. *Plato. Timaeus and Critias*, transl, introd. and appendix, by D. Lee, Penguin, Harmondsworth, England, 1977, 30. E. Hamilton, Huntington Cairns, eds., *The Collected Dialogues of Plato including the Letters*, Bollingen Series IXXXI, Princeton, 1999, *Timaeus*, 34, a, 1164; 38, c, d, 1167.
7. *Plato. Timaeus and Critias*, 42–43, (30–31*)*, *The Collected Dialogues*, 34, b, 1165.
8. *Plato, Timaeus and Critias*, 61–62 (*44–45), Collected Dialogues,* 45, b, c, 1173.
9. *Plato, Tmaeus and Critias,* 122, (*91*) , *Collected Dialogues,* 90–91, e, 1210.
10. *Plato, Timaeus and Critias*, 122 (*91–92*), *Collected Dialogues*, 91, a, b, c, d.
11. B. Russell, *A History of Western Philosophy*, Allan and Unwin, London, 1946, 184.
12. *The Collected Dialogues of Plato*, Laws, X, 886, d, 1442.
13. *Ibidem*, Laws, X, 889, b, c, 1444.
14. *Ibidem*, Laws, X 891–892, 1446–1447. Cf. also: Farrington, *Greek Science*, 36 en W.K. C. Guthrie, *The Greek Philosophers*, 88–89.
15. *The Collected Dialogues of Plato*, Phaedo, 79 c, d, 80 b, 62–63.
16. The question is fully discussed by Gosling in his *Plato* J.C.B. *Plato*, Routlege and Kegan Paul, 1973.
17. *Ibidem*, 31.
18. *Ibidem*, 52.
19. O. Hamelin, *Le Système d'Aristote*, published by L. Robin, Alcan, Paris, 1931.
20. *Ibidem*, 373.
21. *Ibidem*, 382.
22. *Ibidem*.
23. Aristotle, *On the Soul, Parva Naturalia, On Breath*, transl. by W.S. Hett, Loeb Classical Library, W. Heinemann, London, , Harvard University Press, Cambridge, Massachusetts, I964, *On the Soul*, III, V–VI, 430 a 10–25, 171.
24. *Ibidem*, 386.
25. *Ibidem*, 389–392.
26. C. Blakemore, *Mechanics of the Mind*, Cambridge University Press, Cambridge, London, 1977.
27. Farrington, *Greek Science*, 103.
28. A. L. Peck, (transl.), *Aristotle. History of Animals*, Introduction to Books, I–III, Harvard University Press, 1968, I, xii.
29. Part I, Before Philosophy, Chapter 3, i.a. 44–45.
30. Peck, Aristotle, *History of Animals*, I, vii.

31. *Ibidem*, I, xvi.
32. *Ibidem*, I, xxiv–xxv.
33. *Ibidem*, Peck, *History of Animals*, VII, 588 b.
34. *Ibidem*, VII–X, VII, 589 a.
35. *Ibidem*, VII–X, VII, 588 a.
36. *Ibidem*, I–III, II 502 a 16–35, b 1–26.
37. E. Bréhier, *Histoire de la Philosophie. L'Antiquité et le Moyen-Age*, Vol. 2, PUF, Paris, 1955, 450–451.
38. *Ibidem*, 451.
39. G.W.F.Hegel, *Wissenschaft der Logik*, Felix Meiner Verlag, 1963, 145.
40. M. Heidegger, *Was ist Metaphysik ?*, Vittorio Klostermann, Frankfurt A.M., 1949, 7–8.
41. Bréhier, *Histoire de la Philosophie L'Antiquité et le Moyen Age*, Vol. 2, 457.
42. *Ibidem*, 462.
43. *Ibidem*, 464–465.

Chapter 5. The Church Fathers and the philosophers of the Middle Ages: theology and philosophy, a compromise

The religious source of inspiration: Judeo-Christianity

Like in the previous chapter of Part II, my intention is to select a number of elements that have a bearing on the long and complicated historical process resulting in the general western conception of man and of his "anthropina," those qualities that are unique to him, and the conception of his relation to nature and the animals. Of particular interest for our subject are those aspects of this understanding of what it is to be human that are irreconcilable with the modern and contemporary scientific theories. The contrast between science and general culture where our species is concerned will be a major subject of the second volume.

In contemporary Christianity there is a general tendency to concentrate upon the New Testament and to forget as much as possible about the Ancient Testament, because the latter is even more difficult to reconcile with the mentality of our day than the former. However, in order to understand the origin of our conception of the nature of human beings, we must go back to the first, oldest texts of the Bible, which have been and still remain basic for our view of the world and of ourselves. Christianity has determined the picture Westerners have of man in all its fundamental aspects, and this is true as well for the conception of believers as for that of non-believers. Contrary to what could be supposed there is in this respect no essential difference between them. The source is and remains the Old Testament; by the original idea "man" was not revolutionised, only transformed by the New Testament.

We brought already to the foreground the crucial contrast in Greek philosophy between the philosophers of nature and those interested in reality beyond reality. For European culture the Greek naturalistic outlook has been of little or no importance, whilst Greek metaphysics was a major source of inspiration. Our general conception of human beings is determined by the combination of this legacy with Christian theology. It is only in modern times that science slowly but surely developed a view of man that more and more differed from the traditional conception and that had much in common with the view we find in naturalistic philosophies of antiquity.

In the sixth century before Christ, the time of the birth of Greek philosophy, four religions presented rival world views. Three of them were polytheistic, namely Babylonian, Egyptian and Greek religion, whereas the Jews had become over the centuries monotheists.[1] The Egyptians seem to have been rather optimistic about the nature of the world and to have believed that though disasters were always possible, things would inevitably return to their normal course. The Babylonians must have had a more gloomy temperament, as they accepted that things could change more permanently for the worse. The Greeks thought of themselves as distant relatives of the Gods, though the distance between them and the Olympians remained of course immense. Also, the heroes that were half gods constituted a link between the realm of the mortal and the divine beings. In another respect their religion was exceptional too; it was not based upon holy texts, but was rooted in a literary tradition, the epic poems of Homer transmitted through the ages. Though it became a religion of state and the cult of the Gods had to be observed by the citizens, belief in its literal truth was not demanded.

In spite of these differences, all these religions, except that of the Jews, were monophysisms. The Gods were part of the world, they were elements of nature, and for a Greek or an Egyptian for example, seeing the sun rise in the morning was seeing a God begin his daily journey through heaven. Worshipping the sun was the same thing as worshipping Apollo or Ra. The Hebrew religion was not a monophysism: God or Yahweh, was neither a power of nature, nor an ordering principle, but the creator of the world. Asked who he was, he replied "I am who am," which meant that he was pure being and that all further questions would be fruitless. He was understood as totally incommensurable with his creation; he was the source of all being, and therefore Being itself, pure transcendence. God was therefore beyond human comprehension, and the world he created was not the realisation of an eternal rational model, but the result of his free choice. His will became free in an absolute sense, and as a consequence the world and its elements could have been totally different or even not have existed at all, while in the Greek conception the world is not created, but its order is the expression of an eternal divine rationality. Philosophically trained men are able to discover that the world, because it obeys the laws of rationality, cannot but be as it is. The Greeks must have seen themselves and human beings in general either as the *necessary* result of the harmonious construction of the world or as the *outcome of natural processes*, whereas for the Hebrews human beings were freely created by God in his goodness and therefore were *contingent like the rest of the world.*

Old testament

Man owes his life to God

The difference is enormous. In the Greek conception the world is eternal and the question of its contingency makes no sense. Its order is either due to a pre-existing rational plan and men fit perfectly in it, (Pythagoras, Plato, Aristotle) or else it is due to processes obeying natural laws and in that case the human species is a necessary outcome of such a process (the philosophers of nature). In the biblical conception God has chosen at a certain moment, *and for no reason we can understand*, to create this world, and all living beings, including man.

His creation and its different elements could not only have been different if God had willed it, but they could have not existed at all. God could have, after having created man, blown away his most perfect creature from the surface of the earth like a speck of dust. And he could even have caused the end of the world. There is nothing God did he could not undo, which means that nothing exists without God, if not willing it, at least tolerating it. In consequence, not only mankind, but each individual man only exists, because God has allowed or willed it. He decides if a man and a woman will procreate or remain barren. Indeed, at the moment of a child's conception God's breath lends the new individual its principle of life, its soul, as we can read in the Bible: "The Lord God formed a human being from the dust of the ground and breathed into his nostrils the breath of life, so that he became a living creature."[2] This does not only apply to Adam, but to all men; it is the reason human beings cannot dispose of their own life and a mother cannot dispose of the life of her unborn baby. All that exists has the foundation of its existence outside itself, in God. Even today, the believer and the non-believer have a similar general conception of man, but they are distinct in their approach to many ethical questions, the non-believer stressing in contrast with the believer the autonomy of each individual in these matters.

The Old Testament is a rich source for our purpose, but it is very difficult to separate how the texts were intended by their authors from how they were understood later on, first by the Jews, and then by the Christians. For example Yahweh, God, is said to have created the world out of nothing, but this is already an interpretation of what we read in *Genesis*. If it is true that the God of the Bible initially was not a transcendent God, but a force of nature, then the world was created not out of nothing but out of a force of nature. Though the original meaning of the biblical texts and the historical context in which they were written is not our subject, but it is the mainstream Jewish and Christian traditions that formed our conception of man, we must be aware of this.

Man created in God's likeness and appointed the governor of nature

In the book *Genesis* of the Bible, it is told that God created the world and its elements by ordaining that they come into being. First he created the sky and then the earth. Once this was achieved he created light and also day and night, and all this he did on the first day. On the second day, he separated the waters and put the sky between the lower waters and the higher waters. (This was one of God's exploits over which medieval scholars have racked their heads, offering different solutions for this riddle. It seems however, that what was meant with the higher waters was simply the rain falling down and with the lower waters the rivers and the seas). The third day, he separated dry land from the sea and made the earth green with plants. On the fourth day, he made the Sun, the Moon and the stars and separated light and darkness. Also, God populated the seas with fish and other creatures, the sky with birds, and he ordained them to be fertile and to multiply. The fifth day the land animals were made and finally God created man after his image. God speaks, and what he says is realised:

> God created human beings in his own image; in the image of God he created them; male and female he created them. God blessed them and said to
> them, "be fruitful and increase, fill the earth and subdue it, have dominion over the fish in the sea, the birds of the air, and every living thing that moves on the earth." God also said, "throughout the earth I give you all plants that bear seed, and every tree that bears fruit with seed: they shall be yours for food. All green plants I give for food to the wild animals, to all the birds of the air, and to everything that creeps on the earth, every living creature." So it was, and God saw all that he had made, and it was very good....[3]

Fathers and their children resemble each other and Yahweh created man to his own likeness. He puts nature under Adam's and Eve's care, more, he commands them to reign over it, and a sign of the special status he accorded to his human children was that he left to hem the task to give the animals names.

> So from the earth he formed all the wild animals and all the birds of the air, and brought them to man to see what he would call them; whatever man called each living creature, that would be its name....[4]

God made the animals for Adam "as an aid" for him, which means that from the beginning the other species were subservient to man. God does not give to man the order to *exploit* nature, but to *reign over it.* After having placed them in the Garden of Eden, he recommends Adam and Eve to tend it, but he forbids them to eat from the tree of the knowledge of good and evil. As we all know Adam, enticed to it by Eve, bites in an apple from this tree. This is the Original Sin that taints all human beings through the generations.

As can be seen from the fate of Job, God not only demands unquestioning obedience of all his children, but he can also put to the trial a man for no apparent reason. Abraham, willing to follow God's command and to sacrifice his son, has eventually been spared and returns home with Isaac, while Job, who has done no wrong, will lose everything a man can lose. Also, the matter is much more complicated than the sin of revolt: the problem is whether only those who sin and break the law are punished or whether God punishes also good men in order to try them. The answer we receive is at first sight that indeed, when provoked by the Devil, God gives the latter the permission to put to the trial his most perfect servant. He had hitherto favoured him in all aspects, but now he is not spared the most dreadful, the most cruel sorrows. At last, overpowered by his sufferings, Job laments and curses the day he was born. His friends, in keeping with the staunch belief that there is justice in the world and that only sinners are punished, incite him to recognise that he must have sinned, perhaps unknowingly and must repent. This Job refuses; he is innocent and cannot understand that God who is just nevertheless tortures him. This is Job's quandary, but then God speaks to him and asks whether Job can compare himself with the Creator of the world and Job bows his head and regrets his foolish words and lamentations. God then restores Job in his former prosperity. Different answers can be given to the problem of evil in the world, and it seems that here Job comes to the conclusion that he cannot understand the reasons for the interventions of God and to try to do so would be "hubris," a superlative of recklessness. The text we are referring to must probably be understood as the consequence of the cosmopolitanism of the period it was written. The question it raises is presumably a sign of the influence of Greek scepticism and moral relativism.[5]

The problem of evil in the world is totally different for the believer and the non-believer. The latter knows that to ask *why* there is evil and suffering in the world cannot be answered, but that there certainly are natural causes for it. Wars are caused by men, and produce in their turn all sorts of misery; epidemics are caused by germs; illness, physical or mental, can be explained by a great variety of factors, etc. Non-believers content themselves with understanding *how* things came to be as they are, and their only hope is that they can try to change the world for the better. To the believer this is not enough, because he knows that behind these natural causes ultimately there is God, who at least allows evil and suffering, and suffering, not only of sinners, but of men and woman who personally did no wrong, and even of little children and babies. We are not in this world to understand, but to love and serve our creator. The only solution is resignation; humans must bow their head before God Almighty, like Job.

The Hebrew spiritual authorities resisted for a long time the temptation of Greek intellectualism, but it was extremely difficult to avoid contamination of Judaism with the predominant Hellenistic culture. Under the influence of Hellenistic philosophy, mainly of Platonism and Neo-Platonism, a dualistic interpretation of Jewish religion gained in importance. Reality came to be considered as consisting of two distinct realms, the one spiritual the other material. Man became anti-nature in this conception. By being created in the image of God, he contained a divine spark; he was at once a material and a spiritual being and thereby partly transcended nature and was opposed to it. The natural part in him, his instincts and his sensuous nature, had to be tamed.

It is easy to see that though eventually the influence of Greek philosophy could not be resisted, at the same time it remained in many of its aspects alien to the Hebrew world view. The main obstacle for the integration was that for the Jews Yahweh, though he reveals himself occasionally in this world and intervenes in it, is not part of it. The Demiurge of Plato, by contrast, is in this world that he ordered from the inside into a harmonious whole. In the book Genesis it is said that God has created nature and therefore it could not but be good, and he has created man, body and soul and as part of God's work both were sanctified.

In the idealistic interpretation nature and the world of the senses became the lowest form of reality, and the aim of human existence was to reach for a higher reality that could only be attained by cultivating spirituality.

New Testament

Orthodox Christian doctrine versus gnostic belief

Christians have understood the meaning of the arrival of Jesus Christ as an event marking a historical cleavage between the Kingdom of the earthly Jerusalem and the heavenly Jerusalem. It is only against the background of the Hellenistic culture that we can understand the teaching of Jesus and its different interpretations. Not one, but two very different traditions developed and they existed for a long time side by side. By contrasting them we shall be able to grasp the elements of importance for our subject.

It was natural that after the death of Jesus his twelve apostles became authorities on the ground that they had been privileged witnesses of the most important part of his life. Nevertheless, many other men and women had known him and could claim to be his true disciples. Therefore an additional argument for authority was needed, namely marks of the favour of the Master. After his death and resurrection he appeared to the twelve apostles and even lived among them for a certain time. The question was to whom he had appeared first. Jesus must have shown himself to a number of his followers, for example, though not in his earthly form, to two disciples on the road to Emmaus. The four gospels of the New Testament are not unanimous concerning this question. The gospels of Mark and John tell us that it was to Mary Magdalene that he first appeared. In Jerusalem, James, Jesus' brother, became a rival to the recognised authority of Peter, and it was said by his followers that it was he who had been the first to see the resurrected Lord. Visions and apparitions did not seem totally uncommon to Jesus' contemporaries, but what had happened here was a bodily resurrection; he could not only be seen and heard, but also touched, and this was truly a miraculous event. Nevertheless, both Mark and Luke declared that Jesus himself had named Peter as his successor, and John also testified that Jesus had said that after his death Peter had to become the shepherd of his flock. Peter took indeed the leadership of the orphaned apostles and the orthodox tradition confirmed that Jesus had chosen him as the rock on which his "church" was to be built.[6]

The New Testament gospels all are narrative; they relate the birth, the childhood (John), the public life, the martyrdom, the resurrection and ascension of Jesus. The Gnostic gospels discovered at Nag Hamadi, going back to first century traditions, and written by members of groups of followers of Jesus are totally different. They show us Jesus as a teacher and inform us about his sayings, rather than about his biography. It was also discovered, that two of the four orthodox gospels must have had a common source, named the Q-gospel, which like the Gnostic gospels was centred upon Jesus' teachings. Jesus is here represented as a spiritual leader and a philosopher, inspired by cynicism. This could account for the fact that many of the sayings in the gospels of Matthew and Luke do not fit the traditional picture of the Saviour; probably they go back to this lost gospel. Confronting with one another the orthodox gospels on the one hand and the Q-gospel on the other hand, we discover in the first an apocalyptic hero, in the second a wise man, offering an alternative way of life.[7] This explains that after the death of Jesus, parallel to each other different traditions developed. The contrast between the orthodox gospels and the Gnostic gospels is extraordinary. Let us sum up the most conspicuous differences.

As we have seen, in the New Testament gospels the martyrdom of Jesus and his bodily resurrection are the climax that is presented as a unique historical event. The apostles and their successors have, by their presence at the crucial moments in the existence of Jesus, the authority that has been transmitted to the successive leaders of the Church. From the beginning they have offered a number of articles of faith to the believers. The Christian doctrine is simple and can be understood by everyone. Eternal life is promised to those who tread in the footsteps of Jesus Christ following the instructions of the Church. The body is created by God and therefore sanctified, and though the blessed ones will live in spiritual community with God after death, they will only know the full happiness of life in heaven at the end of times, when the Kingdom of God will be realised. Then their bodies will be resurrected and their souls will be reunited with them. Because the body is not considered as evil in se, in the orthodox tradition spirituality is not all that counts, and daily life is not rejected as unimportant. On the contrary its most significant moments are consecrated by rites and the Church offers guidance for all its aspects. The accent lies on the communal dimension of human existence and the worthiness of the human person.

Orthodox Christianity remains paternalistic, and in consequence women can play no role of significance in the Church. Thus the attitude of Saint Paul was ambiguous He seems namely to have accepted that in principle males and females in faith are equals, but nevertheless he stressed the inferiority of women. Commenting on the rules of behaviour in public worship, he justified this as follows:

> A man must not cover his head, because man is the image of God, and the mirror of his glory, whereas woman reflects the glory of man. For man did not originally spring from woman, but woman was made out of man; and man was not created for woman's sake, but woman for the sake of man; and therefore a woman must have the sign of her authority on her head, out of regard for the angels. Yet, in the Lord's fellowship woman is as essential to man as man to woman. If woman was made out of man, it is through woman that man now comes, and God is the source of all.[8]

Following the orthodox doctrine the one God is a man and this has important consequences for the social practice. In contrast with this, in Gnostic texts God is not the creator of the universe, but the creator is his mother, Wisdom. There are different explanations of how God came to know this, but anyway it came as a shock to him and this is how Justinus explains the saying "The fear of the Lord is the beginning of Wisdom."[9] This difference in doctrine is clearly reflected in social practice. In general in the Gnostic congregations the Jewish tradition, which excluded women from public and religious functions and required from them to be submissive to men, was not followed. Women were considered to be equal to men and could put their talents to good use and exert religious functions; they could become priest, deacon or even bishop. This scandalised the orthodox Church Fathers, who castigated these women for their arrogance.

The Gnosis contains a quite different world view from that of what was to become the mainstream tradition. Like in the ancient myth of creation God is male and female. The Creator is the son of Wisdom, i.e. of the Serpent or Earth. Jesus Christ is not identified with the God of Genesis, as in the orthodox conception of the Holy Trinity, but he replaces him. Yahweh has made a world in which both evil and good are present. With Jesus the promise of a better world is fulfilled. The believer must turn his back on the material reality in which good and evil are mixed; the body that is the source of suffering must be mistrusted. The human mind is a spark from the mind of God, and therefore men can find the Truth for themselves and in themselves. In order to reach Truth, we must turn our mind away from our body, its needs are unimportant; the social conventions are to be ignored. We do not sin by our acts, but by our ignorance that is the true Original Sin. The gift of man to find the Truth in an individual spiritual quest, offers him the possibility to discover God in his own mind, and therefore man is the maker of God. The divine spark in us recovers its true nature, which it had lost when descending in a body. By self-discipline and special techniques mystical experiences are possible. From all this it becomes obvious that the Gnostic movement can by its very nature only be attractive to the few. Its stress upon the individual spiritual experience as the true source of happiness excludes the possibility of an orthodox doctrine with a specific content that is passed on from generation to generation.

Joseph Campbell summarises as follows the profound difference between what was to become the orthodox Christian myth created by Jesus' most intimate followers and the Gnostic belief stamped as heretic by the former:

> Nor did the authorized Christian stream and its underground gnostic complement ever really fuse; for in the orthodox tradition the corruption of nature was attributed to man's Fall, but in Gnosticism to the Creator, so that whereas in the former redemption was to be gained through an act of repentance and thereafter obedience to what were taken to be God's laws, the Gnostics strove for release from corruption through a systematic *diso-*

> bedience of those laws in either of two ways, through asceticism or its opposite, the orgy.[10]

(The orgy refers to Gnostic sects where so-called Love Feasts were held that instead of celebrating Christian love among men by sharing food, were festivities showing a strong resemblance to the Dionysian orgy and in which the serpent played a focal role.)[11]

The early Christian conception of man

The orthodox Christians did not believe the body was an element of corruption in se, but only because of the original sin that prevented man from then on and for all times to rule freely over it. In spite of this, the promise of salvation and life after death turned the soul into what is incomparably more important than the flesh. In Plato's time, as we have seen, for the majority of the Greeks it was not at all evident that the human soul was immortal. Nevertheless, even then fear for what came after death and superstition must have been common. Indeed, already in the classical period Democritus found it necessary to reassure his contemporaries that they must not fear death, because when we die our soul, which is a combination of atoms just like our body, is dissolved together with it. We cannot experience our own death and there is no afterlife. Epicurus much later, Hellenistic times, adopted in broad lines Democritus' physical theory, and likewise saw in it a means to dispel the fears and superstitions of those who were attracted to cults promising an afterlife of one kind or another. The Jews originally did not believe in eternal life. To them the soul was the life-giving breath, a gift of God. They were mortal sinners, but at the moment of death they returned their soul to their Creator, which whatever they had done had remained pure as such. If they lived the good life they did so to please God, not to gain eternal life. Their belief in the coming of the Anointed One, the Messiah, in some future time, did imply the final judgement of mankind and the resurrection of the dead, but only after the Kingdom of God would be established on earth. The Christians, rightly or wrongly, have interpreted the death of Jesus as a redemption. Jesus brought in his preaching: "repent and believe in the good news" the message foretold in Isaiah (61:1) that the Messiah would proclaim "good tidings to the poor," meaning that their liberation was near. The Christians, however, believed that salvation meant that the blessed ones would go to heaven, those who are damned to hell, but this is only possible if the soul is not merely the breath of life, but an eternal substance.

From an anthropological point of view this makes all the difference: from now on man exists only during his earthly life as a combination of mind and body, but when he dies and ceases to breathe he leaves his body behind him, but not his true self. This conviction is fundamental for the Western conception of true humanity. To Jews and Christians alike it is clear that God can look into our "heart and kidneys." Only, Christians are aware that we are no longer only judged by our deeds, but by the purity of our soul we have permanently to guard. Sin may be material in a sense, it may come to a person as an exterior force, comparable to an illness, but there is more; the pious man has to be good and obedient in his heart. Man possesses free will and can choose, but God is not content to observe what a man does, but knows also all his thoughts and his secret desires. We are reminded of this by the picture of the all-seeing eye of the Holy Trinity. This is a terrible thought; God being omnipresent and knowing all, we cannot hide from him and we can have no secret for him. Nobody is without evil and even if we are not aware of having done anything wrong, we have committed at least daily, i. e. minor, sins.

If for the Christian our eternal soul has become our true self, *self-consciousness* is all-important. To be a person is not only to be able to weigh good against bad and to choose to be obedient or disobedient, but to be capable of self-examination, to reflect upon oneself and decide what were one's *intentions*. Reading the famous analysis by Kierkegaerd of Abraham's trial when he is asked by Yahweh to kill his son, we are confronted with a series of possible mental scenarios. The facts remain the same, and we will never now what went on in the mind of Abraham when he was ready to thrust the knife into the heart of his son, but to Kierkegaerd the drama is exclusively internal and he imagines this internal turmoil. From his analysis it becomes clear that the testing of Abraham that seems so cruel must be understood as the offering by God of the possibility of an "existential leap" by which Abraham proves that his faith is not merely admirable, but absolute: *credo quia absurdum est.*

From early on the idea that man is *essentially* a subject, that he *is* his consciousness, but *has* a body becomes evident to the Christian Westerner. The unavoidable Hellenisation of Jewish culture explains that dualism became part of the Hebrew anthropology, though this was originally foreign to the Jewish view. Originally it was alien to the Christian view too, and as we shall see, the belief in an eternal soul was not unproblematic for theologians. Saint Paul did distinguish in man two natures, the first one soiled with the Original Sin, the second still pure. Nevertheless, he did not interpret body and soul as two different substances like Plato, associating temptation and sin with materiality and innocence with spirituality. Anyway, Paul's vision was of no great help for understanding the relation of body and soul, which will remain mysterious throughout the Middle Ages.

Another element that is important for our investigation is the concept of *time*. Time in the Bible is historical, all events are located in time, to begin with the Creation. First God makes the earth and the sky, but all is still dark, then he creates light, separates it from darkness and calls former "day," the latter "night." He creates the Sun and the Moon and by giving movement to the Sun time starts running. Creation took seven days. From now on we can speak of God before he created the world and after he created the world. The following major event is the Fall of Man, and here again we can see that nothing remains the same after it took place. Time is irreversible, we are historical beings and try as we may, we never can recover our lost innocence. We are forever chased from Paradise: however strong is our longing to return to it, Cherubs guard Eden with the flame of the flashing sword. Man has become a sinner who does not shun to kill his own brother, as we know from the story of Cain and Abel. Yahweh witnessing the depravation of men, curses them and with them all the living beings he created. When God destroys life on earth by the Flood, only Noah and his wife, their sons and daughters in law are spared and of each kind of creature a pair. Again there is before and after the Flood, before and after God's forgiveness and the Ark of the Covenant, and this holds for all the great events recounted in the Old Testament and in the New Testament as well. Time is sacred time for the believer, whether Jew or Christian. To the latter time is divided between before the death of Jesus Christ and the promise of Redemption, and after it. Like the Jew he has faith that the future will bring the Kingdom of God on Earth. Jesus has set an example for all Christians by his martyrdom in which from the evil done to him and from his agony good has come. *Suffering* is part of human existence and must be accepted with resignation, because, as the Lord permits it, it certainly has a meaning that is hidden to us. The hope of the Christian is that he will be recompensed in heaven for having accepted his burdens in this world.

In this religious view man is a historical being in a double sense; he takes part in both profane and sacred history. Sacred history places profane historical events in the perspective of the history of Salvation: the Fall of Adam, the depravation of man, the promise of redemption by the Passion of Christ, who is the incarnation of God on earth, the reaching of the *Eschaton*, the ultimate moment of the final reconciliation of man and his Creator at the end of times. Löwith explains that from a Christian point of view we live in an interim:

> In this theological perspective, the core of history must be understood as a vast movement from alienation towards reconciliation, one long round about way that leads via countless repeated acts of insurrection and submission to the starting point. Only the sin of Man and the intention of God to redeem him, demand and justify historical time. Without the Fall and the ultimate salvation the interim would be superfluous.[12]

It is through the sum of the acts of human individuals that this reconciliation will finally be realised. Each of us contributes by the decisions he makes in a profane context to the sacred history and to the slow process of perfection of the *not yet accomplished reality* created by God. The promise is fulfilled by the coming of Christ and yet not totally fulfilled, man still possesses an ultimate hope, the hope of Heaven on Earth, and this implies that *mankind has a mission* in bringing the happy outcome nearer by.

It is easy to understand that the orthodox Christians could propagate their belief. The Christian apostles founded communities in many parts of the Roman Empire. They were able to attract not only simple people of the lower classes who hoped for a better life after death, but they could eventually gain also influence in higher circles. Two major problems had to be solved by the early Church Fathers: they had to spread the Glad Tidings and they had to guard the Christian doctrine from impure and disparate interpretations. In distinction to the Jews, who in general did not want to convert foreigners to their religion, the Christian apostles were universalists and wanted an expansion of their religion to other countries. It was not difficult to gain adherents in the lower classes of the Roman Empire, amongst the slaves, the craftsmen, the shopkeepers, but it was much more difficult to convince the patricians and noblemen of the important towns. The latter were in their majority cultivated people, instructed in the liberal arts, skilled in rhetoric and logic, and in Greek philosophy. It were these people who were of importance, because they were powerful enough to contribute to the establishment of Christianity as a recognised religion. The example of Saint Paul is well-known. When preaching the Gospel during his apostolic journeys, he was more than once laughed at or even molested. Like the faith of the Jews the faith of the Christians was simple, it spoke to the heart, more than to the rational mind. Though Paul was a Roman citizen and had notions of Greek philosophy, and though he applied the usual rhetorical conventions when addressing a cultivated audience, he could not convince his public. To educated people Christian faith was a special brand of the Jewish religion. It contained ideas and followed a logic totally alien to people imbibed with the basic ideas common in Greek culture. If the Christian leaders did not want to be considered as simple and credulous men belonging to a fanatical sect, they had to adjust their doctrine to the intellectual standards of the people they wanted to persuade. They succeeded to give their faith the necessary rational backbone. Finally, in the early fourth century, under Constantine, Christian religion became state religion.

The Church Fathers and the philosophers of the Middle Ages

Religion and philosophy

Even Israelite religion had not escaped the influence of Hellenism. Amongst Jews there had been a great resistance against the influence of Greek culture, and especially against Greek philosophy. The Jews were very conservative in matters of faith and ready to go through great lengths to defend their articles of faith. Nevertheless, the introduction of Hellenism could not be stopped and a synthesis was sought between Judaism and Platonism. Moreover, society had changed; a more humane approach to social problems was adopted and this required certain adaptations on the cultural level too. Yahweh, the stern and revengeful Father, whose decisions could not be understood by human beings, had become more and more a moral and reasonable God. Philo the Jew was able to realise the integration of Platonistic elements in the Jewish doctrine. In the new conception, though God had created the world out of nothing, he had, like Plato's Demiurge, also ordered the world in a rational way.

Thus Yahweh, a long time before Christians entered the scene, had not only become a morally good God, but also a rational God. The original Christian faith, however, was not the faith of intellectuals but of simple people, adhering to it with the heart more than by their reason. Saint Paul is clear about this: rationality must be subservient to faith. We can read in Corinthians:

> Jews demand signs, Greeks look for wisdom, but we proclaim Jesus nailed to the cross; and though this is an offence to Jews and folly to the gentiles, yet to those who are called, Jews and Greeks alike, he is the power of God and the wisdom of God.
>
> The folly of God is wiser than human wisdom, and the weakness of God stronger than human strength.[13]

All the same, in the Gospel of Saint John the influence of Greek philosophy is evident. Christianity is wavering between two standpoints: on the one hand faith is sufficient, it is an "*élan du Coeur*" so strong that it makes rational arguments superfluous, on the other hand, because God is by excellence rational, to believe equals understanding. This ambivalence in the positions is of great importance, because he who surrenders himself with body and soul to God without demanding reasons for doing so, and he who cannot love God without knowing why it is that he loves, are two different persons. It is clear, however, that both *religiosity* and *rationality* are on the list of the unique qualities of man, though these elements did not have at all times the same importance for the conception we had of what a human being really is.

Judaism eventually had not succeeded in warding off the influence of Hellenism, simply because Hellenism had become the dominant culture and the pressure became too great. For the Christian leaders who were inclined to proselytism and had to preach for people impregnated with Greek culture, it was even more necessary to consider the possibility to adopt parts of Greek philosophy in order to make Christian religion more palatable to pagans. Nevertheless, this did not always go without difficulty as there was internal resistance against it. *Grosso modo* three attitudes were possible. The first was to reject philosophy altogether as foreign to and even dangerous for Christian faith. The second option, mainly proposed by Church Fathers who had themselves been educated in the Hellenistic tradition, was to acknowledge the possible contribution of philosophy to the development of Christian doctrine. The third attitude was adopted by the Fathers influenced by Gnosticism, which contained, as we have seen, a mixture of pagan philosophy and oriental mythical elements. They were of the opinion that for understanding the mysteries contained in Christian religion one had first to acquire a corpus of appropriate knowledge and could only then reach the Truth of Faith. The latter view was most severely condemned by the majority of the Fathers, because it implied that Christian doctrine was the outcome of a philosophical search and not a revealed truth. The orthodox view was that philosophy could only be used as a tool in the construction of a unified theological system that would be in accordance with the main tenets of the cultural climate of the time. The primacy of faith over reason had to be respected, and it is in the light of faith that the philosophical notions and theories had to be interpreted and not vice versa. The core of the doctrine had to remain understandable for simple people, but its more elaborated form had to be intellectually satisfactory.

It stands to reason that to reach this goal a careful selection had to be made amongst the different philosophies that were fashionable in the first century A.D., and it goes also without saying that no philosophy containing a naturalistic cosmology, a naturalistic approach of nature and of man could be retained. It was immediately clear that the old philosophies of nature could not be brought into accordance with the articles of faith. The atomists, Democritus and Epicurus became the special enemies of Christian religion and were immediately excluded. Furthermore, only an ethical theory that recognised the divine origin of moral values could be considered, which disqualified scepticism and the theories of the sophists. Cynicism seemed even less recommendable, as it rejected all social conventions and preached a return to nature. This left as alternatives Platonism, Aristotelism and Stoicism. Aristotle, however, must not have seemed a good choice to the early Fathers, because his view of the world was deterministic and the role of God in it was restricted and in addition the individual soul was perishable. This left Platonism and Neo-Platonism as suitable for the purpose of the Fathers. In addition, though to a lesser degree, some aspects of Stoicism seemed acceptable. It must be said, however, that being pagan after all, these philosophies had to be reinterpreted and brought in accordance with orthodox faith. Where Neo-Platonism is concerned, the main danger was that it could easily and as it were naturally lead the mind of the theologian who was charmed by this world-view to pantheism: God had poured himself out in the world and in consequence could be known in his creation. The orthodox view was and remained that God was transcendent, that he was everywhere and nowhere, not to be fathomed by the imperfect human mind.

In all times Christianity has been attentive to the intellectual tendencies in society. New ideas, new theories, often at first sight totally alien to Christian ideology, have been critically examined, processed, digested and then put to good use. Platonism and Neo-Platonism, duly treated and incorporated, gave Christian doctrine the necessary coherence. In consequence, the Biblical picture of man underwent a number of important transformations, while its fundamental elements could remain unchanged. Amongst the main shifts were, as we have seen, the introduction of a dualistic conception of man: he became a soul plus a body, and the relation between both was not at all clear. Moreover, under the influence of Platonism, each man was at once seen as an instantiation of the idea "man" and as a person. What a man is and what he becomes is determined by the particular soul, form or essence he received from God. It was of great importance that this soul was personal, because otherwise the merit or demerit of the existence led by the believer could have no consequences for his salvation. The gnostic conception that the soul was a spark of the mind of God was to be rejected, because it meant that man contained a divine part in himself that he could autonomously discover, thereby becoming himself God, which was a blasphemous idea.

We noted that the conception of man of orthodox Jews and the dissident Christians was not very different in its fundamental characteristics, because in both views the main source of inspiration is the Old Testament. New is that Jesus preached that we shall love our neighbour like ourselves, that he preached forgiveness and humility. The most important change is centred on the belief in Salvation. This belief makes the Christian outlook upon life very different from that of the Jews, who held that life came to an end with death, when a man gave back the life-giving principle to God from whom he received it. Only from the second century B.C. on did they believe in bodily resurrection, more especially for the righteous, when the Kingdom of God on earth would be established. In their opinion their merit when they live the good life, obeying the Laws of God and his orders without the hope of a reward after death, is superior to that of the Christians. The latter live in the hope of the salvation of their soul and an afterlife, because they believe the first stage of this salvation had already been reached with the death and resurrection of Jesus Christ. This preoccupation with personal salvation is a distinctive characteristic of the Christian conception of human existence. Men must always keep in mind their mortality and must prepare themselves to be judged at the moment of death. The soul and its destiny become the central preoccupation.

Saint Augustine. God and the Soul, and nothing else

Saint Augustine is one of the most important pillars of the Church. His work reflects the Platonistic tradition common in his time, but it seems he did not study the *Timaeus*, though a partial Latin translation by Cicero and fourth century commentaries Cassidius were available. The other important influence he underwent came from Neo-Platonism, more especially from Plotinus of which he read a number of *Enneads*. (Augustine and Dionysius the Aeropagite passed the world view it contained on to medieval philosophy, on which it had a lasting influence). Of course, for his aim, which was a coherent belief strengthened by philosophy, he had first to bring Platonism and Neo-Platonism in line with Christian doctrine.

According to him the subject of knowledge is twofold: it is God and the Soul and nothing else is of importance. Basing himself on a letter said to be written by Plotinus, but probably inauthentic, in which the world view of this philosopher is summarised, he identifies the One with the Father, the intelligibles or the Logos with the Son and the World Soul with the Holy Spirit. God is the Holy Trinity, perfect, eternal and unchanging. Direct knowledge of God is impossible; even the categories, the most general of all the concepts, are not applicable to him. The best way to come to an insight in what he might be, is by reflecting on what he is not. Augustine did not conceive of creation as an emanation like the Neo-Platonists, because this was a necessary process, but as a free creation of the world out of nothing. He stresses that, though in the Bible God's work is described as taking place during six days, it has to be understood as a timeless event in which all there is at once came into being. After his conversion, he had rejected the Manicheanism he had once in his younger years preferred above the Christian doctrine, which he had considered to be illogical, nevertheless the problem of Evil kept troubling him. He came to the conclusion that Evil had, in contrast with the Manichean doctrine, not to be conceived as a separate principle opposed to the Good, but the world being a divine creation and the expression of measure, form and order, Evil must be simply the absence of these. Matter is not bad as such, because it contains reason under a seminal form. God, or the One, is also Number, and he has ordered the world following its laws. This is the reason matter can be thought and has, in spite of all, a spiritual aspect.

The noblest creature in the hierarchical order of reality is man. Though he believed that we all must devote our existence to the salvation of our soul, Saint Augustine did not have a pessimistic view of mankind. On the contrary, our nature is good, and it is only through the Fall that we have lost our innocence and are no longer capable of subduing our body as we should. Though it is evident that the soul is superior to the body and its senses, as the beasts have the latter without the former, a human being is the totality of both, and the body is only evil if it is not kept in check. We have the consolation that Evil potentially contains the Good. The best proof of this is that God did not prevent Adam and Eve from sinning, because in his wisdom he knew that from their sin a greater good would come, to wit the possibility of salvation. This idea that from what is evil good can come is linked to the conception of reality as a historical process, guided by Divine Providence. This is a belief that can help us to bear our sufferings with resignation and with the hope of a better world to come.

Augustine did not reject the Aristotelian idea that the soul is a life-giving principle. It contains our memory, our understanding and our will. These three faculties are like the Holy Trinity, they are one essentially and three relatively. But to him, the soul was even more than that, the soul of each person was his *self*. This will have most serious consequences; it has become a dogma, a truth that must not be questioned, or rather something that is evident for everybody.

Westerners see themselves as subjects; it must again be stressed that they *are* their souls, but they *have* a body. It is only in the previous century that neurobiologists began to doubt this. However, the idea that we are our body and the functioning of its organs, be it the functioning of the heart and the lungs, or the functioning of the brain, is after more than a millennium and a half still an idea that is foreign to most people. They share Saint Augustine's opinion that what we truly are is our "self," our "ego." Augustine denies that the soul is a corporeal substance ruling over the different parts of the body and located in the brain, the heart or the blood. Contrary to the view of the atomists, it is not a substance composed of minute individual corpuscles. It is not even our mind and its content, neither the images it gathers from the outside world through the senses, nor the knowledge it "invents," i.e., creates from within. No, this would be to confound the soul with that what it is directed upon, what is outside itself:

> Therefore, let the mind become acquainted with itself, and not seek itself as if it were absent; but fix upon itself the act of [voluntary] attention, by which it was wandering among other things, and let it think of itself. So it will see that at no time did it ever not know itself; but by loving another together with itself it has confounded itself with it, and in some sense has grown one with it. And so, while it embraces diverse things, as though they were one, it has come to think those things to be one which are diverse.[14]

Therefore the precept "know thyself!" is crucial, but we must not know our self as something that is outside us, like a celestial power, or the will of another man, or as we know our own face, which we can only do by looking in a mirror:

> But when it is said to the mind, Know Thyself, then it knows itself by that very act by which it understands the word "thyself"; and this for no other reason than that it is present to itself. But if it does not understand what is said, then certainly it does not do as it is bid to do. And therefore it is bidden to do that thing which it does do, when it understands that very precept that bids it.[15]

The truth must be found in ourselves as the Gnostics believed, but we cannot find it autonomously, only if God illuminates us and makes things inwardly manifest to us. We have to consider the soul as a substance, but not a corporeal one, despite the fact that for most of us it is difficult to imagine something that can exist on its own and yet has no property we would attribute to a corporeal substance. The soul cannot be located in space; it is a special substance "without extension of place by any length and breadth and height."[16] Like later for Descartes, whom he inspired, the great problem was how body and soul are related. Saint Augustine had admitted that he himself had difficulty to understand this and, while in most cases he offered guidance throughout the Middle Ages in most important matters, here he left others with the same quandary. Awkward questions arose, such as whether the soul will feel the fire of hell in a bodily way, and whether at the moment the soul is separated from the body the dead person can retain his or her fullness of being. Moreover, was the soul of each individual created by God, or was it formed after conception together with the body? In the latter case, how could it not die with the body?[17]

The self of a person is that part of his soul that is pure self-awareness, which is not the same as what we mean by self-consciousness in general. We are of course often aware of how we appear to others, of what we are doing or thinking or saying or feeling, but in the Augustinian sense self-awareness is the immediate, which implies literally without means, or unmediated, grasping of that part in us that remains the same over time and that we recognise as our self. We *have* images in our mind and thoughts, we *have* feelings, emotions, we *do* things, but we *are* neither the images we have, nor our thoughts, nor our feelings, nor our deeds, but the *subject* that imagines, thinks, feels, does and is immediately present to ourselves. From this immediate awareness that we are a subject, it follows that we cannot doubt that we live. The original formulation by Augustine of what was later expressed as "cogito ergo sum" by Descartes, is the following:

> Yet, whoever doubts that he himself lives, and remembers, and understands, and wills, and thinks, and knows, and judges? Seeing that even if he doubts, he lives; if he doubts, he remembers why he doubts; if he doubts, he understands that he doubts; if he doubts he wishes to be certain; if he doubts he thinks; if he doubts he knows that he does not know; if he doubts, he judges that he ought not to assent rashly. Whosoever therefore doubts about anything else, ought not to doubt of all these things, which if they were not, he would not be able to doubt of anything.[18]

God gave us free will, and like all things that stem from God it must be good. This could be doubted by those who reason that it gives us the opportunity to sin. God knows beforehand what an individual will do, but is not responsible for it, as it is the result of a free choice. That it is a good thing to be free, can be shown, because without free choice we would simply not be able to live a moral life. Augustine, like Plato, believes the soul is directed to the objects of its love: it can turn its love towards what is infinitely higher, to God, and sear upwards, or it can be tugged downwards if it loves the wrong things under the dictate of its body. In the first case a man's soul, without becoming God, reaches a divine state, in the second a man becomes more of a beast.

In Saint Augustine's conception of man the seeds are present for millenary western convictions. Human beings are the most noble creation of God. He donated each of them with an eternal soul that should elevate itself by turning towards spirituality, and evade the lust of the senses. Though during his existence on earth he is composed of a soul *and* a body, the relations of which are difficult to explain, his true self is not composed of both, it is not the sum of his various mental faculties, but his inner self. It is that part in him that is incorporeal, without extension, unchanging and eternal. This part we know immediately and most intimately by introspection. It is this self that we mean when we assume responsibility for our acts and say "*I* did it," and "I did it *of my own free will.*" We are intelligent, while the beasts are not, we are self-conscious, the beasts are not, we can feel pain, but they cannot, and this because we have a soul and they have not.

Another element of importance in Augustine's theory of man is the idea that he is the only being contributing actively to history and that he has a special mission. We have seen that the problem of Evil in the world was at the centre of his thought. The work of God is by definition good and man as his creation was originally good too, but he was corrupted by Original Sin. We cannot but be aware that there is much suffering in the world; there is war and illness and poverty. Man, however, is a historical being, contributing actively to the course of events. He writes profane history and can thereby play a part in the construction of the Civitas Terrena, the Earthly State, but the true meaning of his actions must be found in sacral history of which the aim is the construction of the Civitas Dei, the City of God.

Petrus Abelardus. Ethical questions

Next to the preoccupation with the delimitation of the respective domains of belief and rationality, the principal subject for the medieval philosopher was Man in his relationship to God. Ethical questions were all important, because to be certain which choice to make could make the difference between gaining afterlife in heaven or going to hell. In order to know what kinds of ethical problems preoccupied the philosophers, let us for an example turn to Petrus Abelardus. This great logician has given a razor-sharp analysis of different ethical concepts: "imperfection of soul making us liable to sin," "sin itself," namely the will or desire of evil,' "the evil deed."

His exposition is in great lines the following. We all do know that human beings are born with positive and negative qualities of the mind and the heart, of which only some have a bearing upon our conduct, inclining the will to do or not to do what should be done.[19] To have a negative disposition is not a sin in itself, because it gives a person the possibility to resist his inclinations, to triumph over himself in virtue of temperance and thereby win the crown: "It is the overlordship of defects and not physical serfdom which debases the soul."[20] But then, what is sin? Abelardus repeats after Saint Augustine that sin is in fact that we leave undone something we know the Lord demands of us; in this sense it is non-being, a non-reality. This is because: "To sin is to despise the Creator, this is not to do for Him what we should do, or not to renounce what we think should be renounced on His behalf."[21]

But let us not be mistaken: a sin is sometimes committed without an evil will, and therefore the former cannot be defined in terms of the latter. The merit of a pious person is not the same as to have no bad intentions, but rather to triumph over them and make them subservient to the divine will.[22] Indeed, God is more pleased with a man who acts in the right way, even if it costs him a great effort, than with a man who anyway wanted to do what is good. Desire in se is never bad, it is only natural; it is due to the weakness of human nature. Our will, however, is free, and we can resist our instincts and evil emotions. No action is determined and a sin is never a predestined event. It must be remarked, however, that a wish, which we do not wish to have, cannot be called voluntary in the ordinary sense. We can neither change our human nature, nor change our individual nature, but we can *transcend* what is evil in them. He expresses the idea common in the western self-conception that in so far we are objectively determined by these factors we are not free, but that it is a unique human characteristic to be a subject, able by its willpower to overcome what is given by nature.

This is the major theme of Abelardus' ethical theory: desire is normal, no physical delight can be a sin *in se*, but desire and delight only become a sin if they are not resisted. How could it be a sin to enjoy tasty food that we need for our health, or to enjoy matrimonial intercourse? This idea is further illustrated by the thought experiment the philosopher invents:

> For example, if anyone obliged a monk, bound in chains, to lie among women, and the monk by the softness of the couch and by the contact with his fair flatterers is allured into delight, though not into consent, who shall presume to designate guilt the delight which is naturally awakened?[23]

Abelardus concedes, rather reluctantly, to the idea that all human beings, even newborn babies, are stained with *original* sin, though they are innocent of *personal* sin. He explains that Saint Paul claims a married couple, abstaining from sexual intercourse, reaches perfection, even though he also said that intercourse is the lesser wrong for those not strong enough to lead a life of complete abstinence. This means that they have permission for the sexual act, and therefore the man and the woman are not to be considered sinners. If the monk does not commit a crime against God by his involuntary delight, the soul that is pure cannot be soiled by what happens externally in the body.

God judges our actions by our *intentions*, and the same action can be good in one context, while bad in another. The consequences of our deeds are not decisive, we can do wrong even if the result of our acts produces something that is good. He explains:

> Augustine reminds us that in the selfsame action we find God the Father, the Lord Jesus Christ, and also Judas the betrayer. The betrayal of the Son was accomplished by God the Father, and by the Son, and by the betrayer. For "the Father delivered up the Son, and the Son himself" (Rom.viii, 32; Gal. Ii, 22) as the Apostle says, and Judas delivered up his Master. The traitor, therefore, did the same thing as God Himself. But did Judas do an-

> ything good? No. Good certainly came of his act; but his act was not well done, nor was it destined to benefit him.[24]

This leads Abelardus to the problem of Evil in the world. He probably thinks of the book of Job, when he writes:

> Everyone knows that the devil himself does nothing without God's permission, when he either punishes a wicked man according to his deserts, or is allowed to afflict a just man for moral cleansing or for an example of endurance. Since, however, in doing what God permits the devil moves at the spur of his own malice, the power which he has may be called good, or even just, while his will is forever unjust. He receives, that is, the power of God, but his will is of himself.[25]

The same reasoning holds in the case of Abraham. God permits nothing and does not consent to anything "apart from rational cause," as Abelardus puts it. Therefore, according to him, the Lord did not intend Abraham to commit a sin by sacrificing his son, but wanted to test him and to put an example to others. God, when he gave this terrible command, knew Abraham and therefore knew that no evil would come from his command.[26] God is "inspector cordis et renum," and in consequence each individual man must look in his own heart and kidneys, he must practice *introspection*. If he repents an evil deed, he knows God will judge him for his intentions, not for the deed as such. He is not only self-conscious, in the literal sense of actually, empirically knowing about himself, he is also conscious of the content of his consciousness; he has *a meta-consciousness*. Thus again it is stressed man is free because he is *aware of his own thoughts and intentions*. His mind is introspective and this unique human quality will remain a frequently recurrent theme in philosophy.

Thomas Aquinas. Reconciliation of faith and rationality

It could be thought that Thomas Aquinas would propose a totally new view on man as he has the reputation to be the philosopher who *par excellence* reconciled faith and rationality and who succeeded to make broad synthesis between Aristotle and Christian doctrine that would last until well into modern times. His conception of human beings differed indeed considerably from the traditional Platonistic perception of man. Though he successfully introduced many Aristotelian ideas, often after having made them acceptable from a Christian point of view by interpreting them in ingenious ways, he sometimes disagreed with the philosopher and proposed an alternative, theologically more orthodox theory. At a difference with Aristotle, he did not conceive being exclusively as actuality, but believed that God, who had declared "I am who am," was actual being, but, as Creator, contained also potential being. God cannot be known directly, of course, but he can be known through his works. In this way he could justify the naturalism of Aristotle and defend a natural theology: to know nature in its variety is to approach God. Augustinians had always despised the sensorial reality, now it was a sin not to respect all forms of reality, included the realm of the senses and this was revolutionary. Hyman and Walsh, in *Philosophy in the Middle Ages*, comment:

> The world of nature thus takes on a dignity which helps to legitimise the philosophical naturalism of Aristotle. The sensible world is not to be taken as a deficient symbol of a more purely spiritual realm, but as one of the levels of divine creation, with its own genuinely operative causes and powers. Such powers include man's natural reason, which is adequate for the knowledge of natural essences and certain truths deducible from them, such as the existence of God.[27]

However, most revealed truths cannot be derived by natural reason. Did Thomas not declare that "it is better to understand than to believe, if the choice is left to us," and that, in consequence, when the conclusions we draw are in contradiction with faith, reason must subdue to faith?

The problem that faced him was not easy to solve: he had to show that the acceptance of the Aristotelian theory of the soul did not lead inevitably to the unacceptable conclusion that the soul is mortal. Aristotle had in a rather vague manner referred to an intellectual soul that was general and eternal, but this brought no solution, because it was the *personal* soul that survived after death according to Christian doctrine, and of which the quality was determined by the merit of a person. It would be heretical to assume that many individuals could share one soul, "for it would do away with the distinction of rewards a punishment."[28] Therefore: "the multiplicity of the souls is in proportion to the multiplicity of the bodies; and yet, after the dissolution of the bodies, the souls remain multiplied in their being."[29]

Thomas had an original conception of individuation that had consequences for his theory of the soul. Matter is either determined or undetermined. Like all other substances, man is composed of matter and form; indeed the former being undetermined must be complemented by the human form to become an actual man. However, concrete human beings owe their individuating principle to determined matter, "*materia signata*," which means that he or she is not flesh and bones, and blood, etc., but *this* flesh, *these* bones, *this* blood.... This provoked criticism, because it is difficult to understand how, if it is matter that individuates, when the body dies the personality still can survive in the soul that is per definition incorporeal and immaterial. E. Gilson, specialist of Thomism, mentions the problem in his study on the subject, and explains (in an elaborate footnote) that matter can make possible the multiplicity of certain forms, but exists only for the forms, and not for itself. Matter is the passive principle of individuation, the form the active principle of individuation. He puts it thus:

> Matter only individualises the soul under the form of a body, and this body is itself only like it is, because the soul lends it its organisation, life and, in one word, existence. Naively we represent when thinking about ourselves a soul that is apart and a body that is apart, and we are scandalised that such a noble substance as the soul can be individualised by a piece of matter. In fact, the body only exists through the soul, and both only exist by the unity of the existential act that causes them, traverses them, and contains them.[30]

His conception of the soul must be seen in the light of the ontological order in the divine creation. Like Aristotle he regards nature as a hierarchy of forms, a *scala naturae.* Of course inanimate matter is the lowest degree, and above it are all the living creatures and, as their highest form, man. Not surprisingly, the soul is the life-giving principle, but whereas for Aristotle the consequence is that when a creature dies its life-giving principle dissolves, for Thomas it survives because it is incorporeal and immaterial, and hence incorruptible *per definition,* and its immortality has not to be demonstrated. Nevertheless, he goes through great lengths to prove this. An important argument is that substances in contrast with accidents are only corruptible by their own nature, just like they can only be generated in that manner. The soul is a self-subsisting substance, or, put otherwise, it has being in itself and does not owe it to the body, therefore it cannot be corrupted by something that is accidental to it. Another reason to be certain of life after death is that everything that has an intellect always desires to exist. From this he concludes: "But a natural desire cannot be in vain. Therefore every intellectual substance is incorruptible."[31]

It is clear that Thomas has to introduce a proper distinction between man and the animals. He declares:

> Now it was shown above that the souls of brutes are not self-subsistent, whereas the human soul is, so that the souls of brutes are corrupted, when their bodies are corrupted, while the human soul could not be corrupted unless it were corrupt in itself. This is impossible, not only as regards the human soul, but also as regards anything subsistent that is form alone. For it is clear that what belongs to a thing by virtue of the thing itself is inseparable from it. But being belongs to a form, which is an act, and this by virtue of itself [and he has defined the soul as an act]. And thus, matter acquires actual being according as it acquires form; while it is corrupted so far as the form is separate from it. But it is impossible for a form to be separate from itself; and therefore it is impossible for a subsistent form to cease to exist.[32]

It would be blasphemous to accept that the same is true of animals and that their soul too is a pure form. He explains that there is an additional reason for the distinction:

> (…) the saying that man and animals have a like beginning in generation is true of the body; for all animals alike are made of earth. But it is not true of the soul. For while the souls of brutes are produced by some power of the body, the human soul is produced by God.[33]

Moreover, the soul is not one, but it is in accordance with the various degrees of the hierarchy of beings. In humans it is not only a life-giving principle sustaining the vital functions of the body, but more perfect as it also performs cognitive operations, i.e., it is also a rational soul.

Nor should we forget that the soul is such that it can only perform its functions when linked to a body, but not tie it too tightly to the body. The human soul is not kindred in nature with that of the beasts, but it is also not equal to the pure intellect of the angels. It cannot know immediately the pure forms, but must derive knowledge from the information the sense organs provide, and reach an intuition of the forms by generalisation. Humility demands that we must recognise this as the limit imposed by God on our intellect. Of all the intellects ours is most remote from the divine mind, which is clear from the fact that our soul can only acquire knowledge if it is united to a body. Yet, of all the creatures with a body and a soul, we are the most perfect, because we can perform cognitive acts in which the body does not participate.

We have seen that even Saint Augustine was puzzled by the relation between body and soul and that, because he did not offer certain guidance in this matter, the question remained unclear for a very long time. By giving too much credit to the Aristotelian thesis that the soul is mortal the Averroists had done nothing to change this situation. Here again we can witness the astonishing inventiveness of Thomas. As we have seen, central in his thought is the creed that everything God has created has to be good. Therefore, he is certain that matter is good, and in consequence he rejects the Platonistic and Neo-Platonistic idea that the soul is punished for having lost its initial purity by being incarcerated in a body. It is a mistake to identify the essence of man with his soul, which is also the error of the gnosis, in which the soul is considered to be a spark of the mind of God. Thomas' position, on the contrary remains in the limits of orthodoxy: the truth is that what is human about man is that he is at the same time a body and a soul.

Thomas Aquinas became in his own lifetime an authority and the broad synthesis he had made was so strong, that in times of crisis the Church called upon him. His ideas about the soul were progressive in his own time. Instead of separating spirituality and materiality he united both in man. Descartes will build a metaphysics where this link between both is cut, and where animals are mere bodies functioning like a machine, while men are defined as bodies plus a soul. This is the thesis that Thomas had, many centuries before, explicitly rejected as a naïve view. The extreme dualism of Descartes was soon recognised, even by other rationalists, to be untenable. The theory of the soul of Thomas is much more in accordance with common sense, yet it remains a form of dualism, and it will prove to be almost ineradicable, not only in theology, but also in philosophy. The mind-body problem is still seriously discussed by cognitive philosophers of the 21st century.

Duns Scotus. Essence and existence

A new question that is important from an anthropological point of view arose in the late Middle Ages, namely the relation between essence and existence in the light of the absolute power and freedom of God. Till the thirteenth century Plato was the greatest authority among all the ancient philosophers, but the influence of Aristotle had grown steadily over the centuries. The universities had become the most important centres of knowledge and all learned people had, before they could study theology, medicine or law, to attend to courses in the "*artes liberales.*" In the artes the main source consulted was Aristotle, whether in logic, physics, astronomy, or biological matters. This was seen as a threat to orthodoxy, *inter alia*, because Aristotle's views about causation were deterministic. In his philosophy God was the guarantee for the order and harmony in the world, the first cause of all things, but once he had set things in motion nothing happened that was not caused naturally. From old the Christian view was that God had not merely ordered a pre-existing world, but had created it of his own free will and could intervene in it as it pleased him. The reaction to Aristotelism was predictable: the traditional idea of the omnipotence of God was now more than ever stressed.

Duns Scotus, dubbed Doctor Subtilis for his very difficult, but indeed subtle style of reasoning, declared that though God certainly was omniscient, his free will had precedence over his intellect, which meant that he could at all times change the course of events. He had not ordered the world like the Plato's Demiurge following the necessities of rationality, but made it out of his free will. He therefore could have made a world different from the one he created, in general lines and in detail. All that exists, except God himself, is contingent, and all that happens is contingent too. This has serious consequences for Scotus' conception of man. God has perfect knowledge of all the ideas, and as a consequence of his own essence containing them. Yet, he does also have knowledge of all the individual things that are their concrete realisations. He states:

> Concerning these questions, the certitude of the divine science regarding everything, pertaining to all conditions of existence, is maintained on the ground of Ideas which are posited in the Divine intellect. And this is because of their perfection in representation, since they represent that whose Ideas they are not only each as a whole in itself, but also according to every characteristic and every situation. And so there is a sufficient reason in the divine intellect, not only for apprehending what has Ideas in an absolute way, but also for apprehending every union of them and their every mode pertaining to their existence.[34]

In other words, God knows all the possible combinations of the properties of individuals and therefore he can choose amongst them those that will be realised.

Though things can be caused by chains of natural causes, ultimately all these causes are caused by God's free will, and moreover, God can, if he wills this, cause them immediately:

> Besides, whatever is produced by posterior causes can be immediately produced by the first, and then it would be contingent just as it is now contingent. It therefore has its contingency even now from the first cause, and not from the proximate cause only. Besides, God produced and produces much immediately, as he created the world and now creates souls, and still everything (is produced) contingently.[35]

Duns Scotus not only declared God has chosen each of us and given each of us existence. This is in accordance with the foregoing, for to exist is to be chosen as a definite combination of properties that form our essence. The ideas of these properties are not contingent, they are perfect and eternal, but their combination is transient. It could be supposed that in consequence there remains no distinction between essence and existence at all, but this is not the case: though not actually, they are *formally* different. Each individual is distinguished from all other individuals by being this particular example of the general sort of thing it is; it is the specific difference that individuates his essence.

William of Ockham. Corrosive logic

William of Ockham went one step further and identified essence and existence: essences exist only as concepts in our minds, they are sets of general properties, and the general has no reality whatsoever, save as the product of our thought. Thus "mankind," and "the species man," do not exist, only this man and that man and still another man... Outside our minds essences and natures exist only *in* concrete individuals, but not independently, as substances to be distinguished from these individuals. He rejects the formal distinction introduced by Scotus, who linked the individual and the universal he considered to be equally real. Universals in Scotus' opinion exist really and distinctly in individuals, though not as the same in each of them, but as differentiated by a "contracting" form. Ockham has no use for such a contracting form that makes the difference: Socrates and Plato are numerically distinct; they are both human, but their humanities have no independent existence, they coincide with their individual existences. Socrates is human in another manner than Plato, and therefore their respective humanities are unique. *Essence and existence coincide*, which is a logical consequence of the principle of the primacy of the individual.

Not only essentialism was thus discarded, Ockham's philosophy was corrosive in other domains too. Duns Scotus had in a way loosened the link between the Creator and his Creation, the latter being no longer the necessary realisation of the content of the Divine Mind. Ockham follows him here and though God knows the world and his creatures, they are by no means necessary, but exist only because he *willed* them. The will of God is absolutely free; will has precedence upon reason and the only thing it cannot do is to realise what is incompatible with the laws of logic. Even if man can participate, though to an imperfect degree, in divine knowledge, he cannot understand the divine will. If all that exists could have been otherwise, man can only bow his head before a transcendent reality to which he owes, like everything else, his existence. Each human being is a divine mystery to himself and to other men.

This can easily lead to a certain form of scepticism, as we must no longer try to understand *the reason why* things are ultimately as they are. Even the essences are secondary to God's intellect and came only into being by God's knowledge of himself. This means that the eternal forms, if not created by God, derive their existence from a special intellectual act of God. The truths of faith cannot be demonstrated and the natural phenomena, being contingent, can no longer be studied by speculative thought about a hierarchy of essences, but have to be known intuitively, i.e. one by one by the senses. Hitherto men conceived of themselves as part of an eternal order of things, from which they derived their properties and capacities. In this way the general structure of which they were elements gave sense and purpose to their lives. Now the absolute freedom and power of God frees them of these determinations, but they are also bereft of their certainties.[36] The world order that was so reassuring, wherein each element had found its necessary place, is lost, and what is left are merely regularities in nature which it pleases God to maintain. Even in the moral domain, if men are no longer the incarnations of an eternal model, they can no longer rely upon preestablished norms: God's pure free will, not eternal laws that would determine what God commands, is what has to be obeyed.

When discussing man, the one being God created in his image, it is on the soul that the debate is centred. Specialists agree that Ockham's views remain as Christian as those of any other Christian thinker of the Middle Ages. What is different is that he explicitly separates the articles of faith that cannot be proven, and arguments that he draws from experience and logic. Thus he does not try to establish the truth he accepts on religious grounds on a metaphysical basis, but only by referring to what we know as facts and what we know by logical reasoning, thereby clearing the path for his successors and preparing future developments. By declaring that reality as a creation of the free will of God is contingent in all its aspects, and that nothing is of necessity as it is, except God himself, he frees the mind of many accepted truths and conceptions.

Where the soul is concerned he applies his usual method that consists in discarding all redundant distinctions and the introduction of corresponding entities that are not necessary for explaining the phenomena under scrutiny. What has to be accepted by the Christian is that the soul is the image of the uncreated Trinity. Furthermore, it is an article of faith that the intellective soul is incorruptible and eternal and that it is the body's form, that the will is free and that the intellect is active. None of these can be proved, however.[37] His own opinion is that the soul is indivisible, and for all we know, that it consists of the emotions, the acts of the will and the cognitive operations of the individual. He identifies the intellective soul with the incorruptible and eternal soul religion demands us to believe in.

Like Duns Scotus he posits the primacy of the will over the intellect and declares that the will is nobler, though they both are aspects of the power of one soul. The habits of the soul are acquired and not innate. They are central in knowledge, they are central in volition. Like Aristotle, Ockham is convinced that the mind is initially a "*tabula rasa*," and that our intellectual habits are written on it by repeated experiences that lead to knowledge, while the habits of the will come about after repeatedly performing an act. Habits in their turn can explain our inclinations, and can explain how it is possible to resist our desires, though they remain present in the senses.[38]

Clearly the matters of the heart, religious beliefs, are separated by him from the matters of the mind, from what we know. He was not, as has often been suggested, a sceptic, but nevertheless, by replacing traditional metaphysical argumentation by what can be known by the senses or by logical reasoning, he encouraged critical reflection. His influence has been important, and it can be said that, although his philosophy was not continued in a school strictly adopting his views, a considerable part of his ideas has been passed on from generation to generation for centuries.[39] There is no doubt that he cleared the path for the progress of modern science and had a great influence on empiricism.

In followers of Ockham, such as Robert Holcot and Adam from Woodham, the scepticism that in the master's work was only in present *in ovo* became more pronounced. They no longer recognise a link between Grace and Salvation; no free will is needed for an act to be judged good or bad; God can love more the sinner than the pious man. God does not know the contingent future, he has a contingent knowledge of contingent things. He can even be mistaken and induce men into error. This scepticism is outspoken in the thought of Nicolas of Autrecourt, who reduced all certainty to probability, with as only exception direct perceptual knowledge.

In fact Duns Scotus, in an attempt to save the Augustinian tradition and to counter the dangers of Thomism, was tempted to introduce a number of novelties that had serious consequences. Ockham studied his theses, commented upon them and drew his own conclusions from them. While being a scholastic philosopher, he undermined scholasticism, and it was impossible to revert to the monumental synthesis between faith and rational thought Thomas Aquinas had realised. Though it will take many centuries before science will free itself totally from the authority of religion, from the fourteenth century on the process is initiated.

The Renaissance

The immortality of the soul in Renaissance philosophy.

In the fourteenth century Ockham introduced the "via nova," and was of great influence, while others, and curiously amongst them Platonists, continued after him what Gordon Leff has called "the dissolution of medieval thought." In Leff's opinion there is no real continuity between the thirteenth and fourteenth century. Perhaps, defining the Renaissance negatively as a mainly cultural, far more than a new social, economical or even political phenomenon, which consists in the rejection of the traditional medieval scholastic approach of man and nature, we could say that the true break must be situated in the fourteenth rather than in the fifteenth century. After the first enthusiasm about the Renaissance as a period literally of rebirth, which was inspired by Jacob Burckhardt, the opinions became more shaded and it was generally accepted that it was at the same time a continuation of the Middle Ages and a genuine new era as well. In the fourteenth century Ockham paved the way for scientific physics, and suggested that another world such as ours was possible. In the fifteenth century, Nicolas of Cusa introduced a totally new picture of the universe. Instead of consisting of a number of spheres, with the earth snugly situated in its centre, he dared to picture it as infinite, with its centre, and in consequence its circumference, everywhere and nowhere. Filippo Lippi, a contemporary of Nicolas of Cusa, painted nevertheless about the same time a wonderfully reassuring cosmology on the cupola of the cathedral of Spoleto. We can admire a world layered like of old, each kind of reality in its natural place and all of them harmoniously related to the encompassing whole. It is clear old and new world views existed simultaneously.

What does this imply for the concept of man? We can without hesitation say that in the Renaissance the picture of the world changed more radically than that of man. Nevertheless, the interpretation of characteristics unique to man and lending "*humanitas*" to all of us, generally recognised by the Schoolmen teaching at Universities, became different. The humanity of men is no longer the sole responsibility of God and a reflection of an eternal model. Individuals are no longer determined by the essence "man," present in the divine mind. On the contrary, each person individually, has to decide what will become of him. Each man, as so often repeated, being a "microcosm in the macrocosm," has gained autonomy and has become for himself the centre of the world. In fact, under the influence of the study of the Ancients, the individual, though owing his existence ultimately to God, marvels at himself. Man has gained a new dignity, *in summa*: he is magnificent. He is part of nature, but also opposed to it; he is able to tame it like he can tame wild animals. It is his prerogative to turn nature into an object for his subjectivity, which means that he can enjoy it, can study it and can dominate it. The individual, full of unique potentialities, has lost the humble

Christian virtuousness and has regained the proud *ἀρετη* of the ancients. He can and must excel by his particular qualities the best he can. He is conscious of his creativity and is confident that he can invent and construct new tools and machines or make marvellous works of art. Thus he resembles more and more his Creator, in autonomy, power, creativity, and though in contrast to what the Greeks believed not being a relative of God, but forever God's modest creature, man is again aware of the divine spark in himself.

Aristotle and Plato remained the most important authorities, but more as a source of inspiration than as proponents of doctrines, which had to be accepted literally. This went together with a remarkable eclecticism leading in some cases to the strangest combinations of elements of *in se* incompatible philosophical, occult and mystic theories. For example it was not unusual, to incorporate elements of astrology and of the Cabbala in philosophical works. Most philosophers still chose between Platonism and Aristotelism, but sometimes we find in their work ideas borrowed from nominalism or Epicurism as well.

One of the obstacles for the acceptance of Aristotle's philosophy had been that he had clearly said that the individual soul was the vital principle of the body and therefore did not survive it. This was of course incompatible with Christian faith. The Philosopher had been less clear about a kind of general soul, sum of individual souls, and this had made possible different interpretations. As we have seen Thomas Aquinas in the thirteenth century finally offered a satisfactory solution for the problem: the soul was the eternal form of the mortal body. However, the debate was not closed; even in the fifteenth and sixteenth century it still lingered on amongst the philosophers of the University of Padova, most of whom were Aristotelians. This university, like others in Northern Italy such as that of Bologna, which was a centre of free-thinkers, was controlled by the authorities of the city and this gave it greater liberty than if it would have been controlled by the Church. A lively argument went on about the survival of the soul after death, because the Aristotelians were divided about the subject and interpreted the Philosopher in different ways. On the one hand the Averroists, like Hieronymus Cardanus contended that the general intellectual soul survives, but the individual soul, linked to the body, is mortal. On the other hand the Alexandrists (followers of Alexander Aphrodisias) believed that, because the intellectual soul cannot function without the images delivered by the sensitive soul that is part of the individual soul, both the general and the individual soul are mortal. The latter theory was held by Pomponazzi and Cremonini and the Pope of that time, Leo X, encouraged Cardanus in his battle against the Alexandrists, being aware that their theory was even more dangerous than Averroism.

Today the Aristotelian philosophers of that time at Padova are mainly remembered for the great liberty they took with religion and in the moral domain. Cardanus who was a pantheist favourable to astrology and occultism, denied like Cremonini that the world was created; Pomponazzi was a libertine, and he denied that the existence of human beings had a purpose outside itself). They were able to contribute to a new intellectual climate, rather than that they were really original thinkers. Now human existence has a worth in itself and not, or at least not only, in the perspective of life after death and of the promise of salvation. If the soul does not survive, only wishful thinking or hope can bring a vague consolation. Life is here and now. Though for these men faith had no longer precedence over rational thought and they proposed unorthodox theories of the soul, and though we can consider them to have stepped outside scholasticism, they contributed far less to the progress of science than the Platonists.[40] Their views were more often than not based upon obsolete, even droll theories, far removed from experimentalism and scientific innovation.

Even if the Church had no direct authority upon the Italian civil universities, several of their professors were condemned for the blasphemies contained in their unorthodox theories.

Notes

1. Cf. R. Kreglinger, *La Religion d'Israel*, M. Lambertin- E. Nourry, Bruxelles/Paris, 1926, 175–190, Cf. also: Albertz, R., *Religionsgeschichte Israels*, Vandenhoeck-Ruprechts, 1992, transl.: A *History of Israelite Religion in the Old Testament Period*, 2 vol., SCM Press, 1994.
2. *The Revised English Bible*, (RVE), Cambridge University Press, Cambridge, 1989, *Genesis*, II, 7.
3. *(RVE)*, *Genesis*, I, 26.
4. *Ibidem*, II, 20.
5. Kreglinger, *La Religion d'Israel*, 312–313.
6. E. Pagels, *The Gnostic Gospels*, Random, Vintage, New York, 1981, 6–9.
7. Burton L. Mack, *The Lost Gospel. The Book of Q and Christian Origins*, Element, Shaftesbury, Dorset, Rockport, Massachusetts, 1993, 30–31.
8. (RVE) 1 *Corinthians*, 11, 7–12.
9. Pagels, *The Gnostic Gospels*, 70.
10. Campbell, *The Masks of God, Creative Mythology*, Arkana, 1991, Penguin, Harmondsworth, 145–146.
11. *Ibidem*, 149–161.
12. K. Löwith, *Meaning in History*, Phoenix Books, University of Chicago Press, 1949, 183–184.
13. (RVE) 1 *Corinthians*, 1, 22–25.
14. H. Hyman, A. Walsh, (eds.), *Philosophy in the Middle Ages*, Augustine, *On the Trinity*, Chapter 8, Hackett, Indianapolis, 1973, 70.
15. *Ibidem*, 70–71.
16. *Ibidem*, 69.
17. G.R. Evans, *Philosophy and Theology in the Middle Ages*, Routledge, London, 1993, 90–92.
18. Hyman, Walsh (eds.), *Philosophy in the Middle Ages*, 72.
19. *Ibidem*,188.
20. *Ibidem*,189.
21. *Ibidem*.

22. *Ibidem*, 191.
23. *Ibidem*, 193.
24. *Ibidem*, 98.
25. *Ibidem*,196.
26. *Ibidem*, 199.
27. *Ibidem*, 464.
28. *Ibidem*, 502.
29. *Ibidem*, 504.
30. E. Gilson, *Le Thomisme*, Librairie Philosophique J. Vrins, Paris, 1948, 267–268.
31. A. Hyman, J. Walsh, *Philosophy in the Middle Ages*, 501.
32. *Ibidem*, 501.
33. *Ibidem*, 502.
34. *Ibidem*, 590–591.
35. *Ibidem*, 592.
36. G. Leff, *Wiliam of Ockham*, Manchester University Press, Manchester, 1975, 311.
37. *Ibidem*, 537.
38. *Ibidem*, 556–557.
39. Cf. J.R. Weinberg, *Abstraction, Relation and Induction*, University of Wisconsin Press, Madison & Milwaukee, 1965, p. 4–5; Hyman, Walsh, *Philosophy in the Middle Ages*, 609; Ph. Boehner, O.F. M., *Ockham, Philosophical Writings*, Hackett, Indianapolis, 1990, Ii.; N. Kretzman, A. Kenny, J. Pinborg (eds.), *The Cambridge History of Later Medieval Philosophy*, Cambridge University Press, Cambridge, 1988.
40. E. Bréhier, *Histoire de la Philosophie*, Tome Premier, PUF, Paris, 1951, 757–758.

Chapter 6. Rationalism versus Empiricism

Descartes' dualistic rationalism at the origin of modern subjectivism

Historical background and sources of Descartes' theory of the soul

In the seventeenth century in France the authorities were even less patient with people expressing unorthodox opinions than in fifteenth- and sixteenth-century Italy. The Jesuits initiated the Counter Reformation with a view to restoring the Roman Church to its former glory in the catholic countries. Probably the Counter Reformation and the influence of the Jesuits upon state affairs explain this less liberal climate. It was necessary for intellectuals to be very careful not to offend the authorities by circulating theories that could sow doubts in the minds of the believers.

It is in the frame of this political and cultural situation that we must understand the intellectual attitude of René Descartes. As a young boy, his father sent him to the Royal College of the Jesuits at La Flèche in Anjou. The Jesuits of this college were certainly influenced by the Thomism of their time that was considered to be the ideological weapon of the Church against the Reformation, but they were perhaps even more attracted to the late version of the Augustinian tradition of the black friars that was fashionable in their own order in Spain and Portugal. The contribution of René Descartes to the change in the perception of man must be understood in this context. Though Thomism was mandatory, the Augustinian tradition was continued at La Flèche, and this feature of his education would become of great importance for his philosophical work later on. Another major source of inspiration was the work of Gassendi under the influence of whom he became an atomist. Descartes resolved the problem of these contradictory influences by the creation of a dualistic metaphysics. Being a quiet man, who cherished above all an undisturbed life with time for reflection, he avoided as much as possible arousing suspicion. It was obvious, however, that he was not an orthodox Thomist, and he never totally succeeded in convincing the authorities of the innocence of his dualistic world view that combined rationalism with dangerous materialism. An eye was kept on him, and he even had to flee to Holland.

Body and soul: two independent realities

Descartes rejected the Aristotelian idea that it is the soul inherent in the body that is its life-giving principle. He had chosen for himself the arduous task to reconcile materialism as a tool for scientific investigations and the spiritualism demanded by religion. Instead of adopting the fashionable Thomist views, he left the beaten paths of Aristotelism. If the soul or "anima" is no longer the source of life in those things that are "animated," what is? Descartes wrongly deduced from phenomena like processes of fermentation of which the cause was not known in his time that Matter, being an active principle, can bring forth life. However, he had still to explain the relation between our mind and the acts of our body, but also the effect of what happens to our body on our mind. Matter is extended, the soul is not, and it has no contact with the body except in the pineal gland, a very small organ at the base of the brain. This is enough to enable the soul to reign over the body, to direct its actions, to guide it by thought and by the decisions of free will. It can hardly be said that this is a convincing solution.

In Descartes' mind it could pose no problem to reduce the body to its purely material aspects, but the consequence of this was that he had to revert to Augustinian doctrine in order to save the intellective soul as a genuine substance with independent existence. Also, he writes:

> I had described, after this, the reasonable soul, and had shown that it cannot be deduced from the potentialities of matter, like the other things I had spoken of, but that it must be expressly created, and how it is not enough that it be lodged in the human body, as a pilot in its boat, except perhaps for moving its members, but that it is necessary that it be linked and united to it, in order to have in addition feelings and appetites similar to ours, and in this way compose a real man.[1]

He refers here to his theory of the pineal gland as the (smallest possible) place of contact, between the soul that has no extension in time and space and the body that does have extension. He continues:

> Furthermore, I have dwelt a little longer on the subject of the soul, as it is one of the most important; because after the error of those who deny [the existence of] God, which I think to have sufficiently refuted above, there is not one that more distances the feeble minds of the right path of virtue than to imagine that the soul of animals could be the same as ours. The consequence of this would be that we have nothing to hope or fear after this life, not anymore than flies and ants, while instead, if one knows how different they are, one understands much better the reasons that prove that the nature of our soul is completely independent of the body, and that in consequence it is not bound to die with it. Moreover, as one does not see other causes that destroy it, one is naturally led to judge from this that it is immortal.[2]

Human beings consist of a body that is a mere automaton, but unlike the body of an animal, it is steered by an immortal soul, which is, to use the words of Augustine, a dweller in a dwelling. It would be a mistake to believe that this dualism was readily accepted; on the contrary, the orthodox position in Descartes' time was that of Thomas Aquinas, namely that the body is pervaded by the soul and that this sanctifies it. It was for this reason that the body must be treated with respect, even once the soul has left it, and that it was forbidden to dissect corpses. This explains why Descartes considered his thesis that the body is a mere automaton that could in principle survive without a soul, to be very progressive. Now the body could be freely studied, dissected if necessary, because it was merely a clockwork or hydraulic system, like those so fashionable in his time at the courts of Europe. It meant that it could be taken apart, its pieces and their mechanical relationships studied, and perhaps one day for the benefit of mankind it would be possible to repair our bodies just like craftsmen repair those devices. It may be so that he contributed by this approach to that analytical spirit that was later on so fruitful in scientific research, but he also strengthened the view that *a human being is a ghost in a body*, that we *are* our minds, but *have* our bodies.

Like Saint Augustine, Descartes believed that all knowledge comes from within. Augustine teaches namely, that the mind contains in potency the ability to know the species; because the mind influences the body, but not vice versa, it is the active intelligence that turns the sensorial data into experiences. Thus the mind must not interrogate the eyes for knowing the truth; they are only the windows through which it sees and it would be preposterous to ask a window what it sees. Descartes' theory of visual perception is that objects reflecting light project an inverse image upon our retina. Our eyes are *camerae obscurae*, dark chambers, in which the material image can be seen by our mind that knows all about the body without being part of it. In contemporary psychology this Cartesian reasoning, which until recently was still present in many theoretical writings, is known as the fallacy of the little man in the brain, because if it is this projected image that is seen by our brain there must be somebody to look at it.[3]

The mind has an activity of its own and can discover the truth without the help of our senses. We can observe and experiment, but only in order to verify what our mind has discovered by reasoning using the concepts God has put in our mind. The question whether knowledge that is not based on or at least controlled by sensorial evidence is not solipsistic simply is not considered. All men must necessarily reach the same conclusions if their reasoning is correct.

Language is a unique human characteristic and without it there is no thought

As we have seen, for Descartes only human beings have a soul. Animals are mere automatons, the lucky ordering of their parts and pieces enabling them to survive and to function. Were it not that God had made them and that therefore they possess a certain perfection of their own kind, nobody would be able to distinguish them from man-made machines.[4] Though our bodies are machines like those of the animals, the difference between a man and a machine is not only due to the number of the parts, the complexity and perfection of their organisation, but is also due to the fact that he can speak. This is not new, from immemorial times language has been considered to be unique to man, but Descartes comments upon the faculty of speech at length and makes some interesting remarks. Indeed, there are no men so backward and stupid, there are no idiots that are not capable of putting together some words and to signify something with them. Machines can be constructed in such a way that when touched in a certain place they produce word-like sounds. Animals too can speak in this sense; they do not lack the necessary organs, as can be seen from magpies and parrots, which are able to pronounce words. Yet, they are unable to use language as we do, expressing our ideas, adapting our speech to the circumstances. Compare this with deaf-mute people who are always capable to invent some signs to make themselves understood by those that are ordinarily with them and had the time to learn their language. We can conclude from this, says Descartes, that the beasts do not have less reason than men, but that they have none. We must not confound words with the natural movements signalling the emotions of animals, nor think, like the Ancients, that animals can speak to each other, but that we do not understand them. It is remarkable, he says, that some animals can do some things better than we do, whereas some of our actions they are not able to perform. This proves they have no intelligence at all, but that it is nature that is at work in them making use of the order of their organs, just like we can see that a clock, consisting of nothing but wheels and cogs, can count the hours and measure time far better than we can do for all our wisdom.[5] The soul, the Ego, self-consciousness were in Descartes' time, and in fact still are, sensitive subject matters, because the existence of something that corresponds to these concepts as an independent substance is the precondition not only for the survival of the soul after death, but for the existence of free will. To give up this prerogative of man seemed blasphemous.

The modern dualistic Cartesian conception became successful; for centuries it has prevailed and even today it has not totally disappeared. Of course it is no longer doubted that to cognitive processes correspond physical phenomena, but the idea that our subjective experiences coincide with natural processes that we can empirically study has remained repulsive to many philosophers and psychologists. It is believed that there must be an irreducible aspect of the human experience that cannot be known empirically and experimentally. Likewise, it has been difficult to convince philosophers, but also linguists, that thought without language is possible. Not until recently was it generally believed that animals are incapable of using a genuine symbolic language. Some still consider animals as a kind of machines, lacking the intellectual ability to form propositional thoughts.[6] Today, finally the influence of Descartes is declining, due to the progress made in empirical research by neurologists of phenomena like consciousness and self-consciousness, intelligence and linguistic skills, and due to new views about the relation between body and mind.[7] Still, only a minority of continental philosophers is prepared to take into account the new scientific theories when treating of the human person. The majority of them take a defensive position and declare that though science is an important aspect of our cultural life, it proposes only a representation of reality, a mere model, a construction that has nothing to do with the intuitive and irreducible personal experience of the subject.

The concept of man of an empiricist: David Hume

Continuity between animals and mankind

For a clear understanding of the matter it is necessary to contrast rationalism with empiricism where man is concerned. It is evident that the empiricists, given the fact that they continue the Aristotelian tradition and consider the sensorial data as the basis of all knowledge and hence posit a well-developed sensorial apparatus as the precondition for intelligence, will view man from a perspective that is totally different from that of the rationalists. The rationalist considers perception at most as a means to control whether theories do not stray too far from common sense. If rationality is the essence of humanity, man will be utterly, and this means ontologically, different from the beasts. Perception, the basis of all knowledge according to empiricists, is on the contrary common to most species, and in consequence there is a continuum between men and the other animals.

Thus Locke, though he was a religious man, and though he believed it was God who created all living beings, is inclined to consider that continuity exists between all species. Certainly, a gradation in perfection is present, and in terms of intellectual power man is exceptionally gifted, but deprived of the acuity of his senses, he would be nothing more than a low-ranking animal on the hierarchical scale. In his *Essay Concerning Human Understanding*, he draws the line of distinction not between men and animals, but between "the animal kingdom and the inferior parts of nature," which is much more accurate, biologically speaking. It is true that some plants, though lacking sense organs, are called sensitive for their brisk motions that show some resemblance to that which in animals follows upon sensation. He explains:

> Yet, I suppose it is all bare mechanism, and no otherwise produced than the turning of a wild oat-beard by the insinuation of the particles of moisture, or the shortening of a rope by the affusion of water. All which is done without any sensation in the subject, or the having or receiving of any *ideas*.[8]

In all animals there is some degree of perception, though in some of them it is extremely dull compared to that of others, and it is proof of the wisdom of the Maker that each animal is provided with the senses it needs. He takes the example of an oyster or a cockle: such an animal cannot transport itself from one place to another, but what good would it do to it, if it could perceive dangerous or desirable objects in the distance, not being able to go in search for them or to avoid them? It has to lie where chance has placed it and receive over it currents of warm or cold, clear or muddy water; therefore it would be an inconvenience to be quick of sensations. Locke's conception of nature is teleological, yet it is clear that the broad ontological division is not drawn between mankind and the animals, but between living beings capable of even the least amount of knowledge of their environment and those without the most elementary form of learning. This must have been a remarkable perspective upon living beings compared to the usual views on the subject in his time. The major standard being sensorial experience, he writes about the lowest-ranking animals and men with impaired sense organs the following:

> But yet I cannot but think there is some small dull perception whereby they [these animals] are distinguished from perfect insensibility. And that this may be so, we have plain instances, even in mankind itself. Take one in whom decrepit old age has blotted out the memory of his past knowledge and clearly wiped out the *ideas* his mind was formerly stored with, and has, by destroying his sight, hearing, and smell quite, and in his taste to a great degree, stopped up almost all the passages for new ones to enter; or, if there are some of the inlets yet half open, the impressions made are scarcely perceived or not at all retained. How far such one (notwithstanding all that is boasted of innate principles) is in his knowledge and intellectual faculties above the condition of a *cockle* or an *oyster*, I leave to be considered. And if a man had passed sixty years in such a state, as it is possible he might, as well as three days, I wonder what difference there would have been, in any intellectual perfections, between him and the lowest degree of animals.[9]

His mind is only at birth a blank slate, but thereafter one that has been written upon by his experiences. However, in old age, most can be wiped out and his senses having become dull offer no longer fresh material. In this philosophy a man is not an immaterial subject, but he is what the state of his body allows him to be. For similar reasons, David Hume never has doubted the continuity between men and animals; the intelligence of animals is attested by all the examples we know from daily life where they show this quality. This is a sound view for treating epistemological questions, because it avoids the very common mistake that is made concerning knowledge, namely to identify it with its most accomplished forms, leaving aside the basic layers common to men and animals. The following text makes this clear:

> Next to the ridicule of denying an evident truth, is that of taking much pains to defend it; and no truth appears to me more evident, than that beasts are endowed with thought and reason as well as men. The arguments are in this case so obvious, that they never escape the most stupid and ignorant.
>
> We are conscious that we ourselves, in adapting means to ends, are guided by reason and design, and that it is not ignorantly nor casually we perform those actions which tend to self-preservation, to obtaining pleasure and avoiding pain. When, therefore, we see other creatures, in millions of instances, perform like actions, and direct them to like ends, all our principles of reason and probability carry us with an invincible force believe the existence of a like cause. It is needless, in my opinion, to illustrate this argument by the enumeration of particulars. The smallest attention will supply us with more than are requisite. The resemblance betwixt actions of animals and those of men is so entire, in this respect, that the very first action of the first animal we shall please to pitch on, will afford us an incontestable argument for the present doctrine.[10]

The philosopher, when writing this, was perhaps thinking of his cat that was such good company for a man with an intellectual style of life, and of which he was very fond. Anyway, the thesis is that an animal functioning accurately in a complex environment must be intelligent.

He draws from the foregoing a further conclusion. If he is right, his theory gives us a good touchstone:

> When any hypothesis, therefore, is advanced to explain a mental operation, which is common to men and beasts, we must apply the same hypothesis to both; and as every true hypothesis will abide this trial, so I may venture to affirm, that no false one will ever be able to endure it. The common defect of those systems, which philosophers have employed to account for the actions of the mind, is, that they suppose such a subtilty and refinement of thought, as not only exceeds the capacity of mere animals, but even of children and the common people in our own species, who are, notwithstanding, susceptible of the same emotions and affections as persons of the most accomplished genius and understanding. Such a subtilty is a clear proof of the falsehood, as the contrary simplicity of the truth, of any system.[11]

Not only is the continuity between men and animals asserted in the cognitive domain, but also to a certain degree in the moral domain. This was an even more progressive idea: the awareness of what is moral is still today generally considered the prerogative of man. The discussion of this matter in ethology, by Konrad Lorenz and others, and its more recent elaboration by Frans De Waal and others, however, offers arguments in favour of his thesis. To begin with, Hume contends that love and hatred, i.e. positive and negative emotions, can be observed in animals:

> But to pass from the passions of love and hatred, and from their mixture and compositions, as they appear in man to the same affections as they display themselves in brutes, we may observe that love and hatred are common to the whole sensitive creation, but likewise that their causes, as above explained are of so simple a nature that they may easily be supposed to operate on mere animals....[12]

Hume notices that animals not only show affection for their congeners, but for members of other species as well, as can be seen from the affection of the dog for its master, an affection that is in many cases returned, and also from the affectionate relation of animals of different species kept together. Of the latter he gives as an example that an ox sharing a meadow with horses will seek their company, though if given the choice he will prefer the nearness of other bovines. Emotions are, after all, more primitive than abstract thought. The instinct in animals to show affection for their young is the same as the instinct in human beings who care for their children. It is evident, he says,

> that *sympathy*, or the communication of passions, takes place among animals, no less than among men. Fear, anger, courage and other affections, are frequently communicated from one animal to another, without their knowledge of that cause which produced the original passion. Grief likewise is received by sympathy, and produces almost all the same consequences, and excites the same emotions, as in our species.....[13]

Today we would use the term "empathy" instead of sympathy, referring to the capacity of social animals to understand the emotions of other individuals in certain circumstances. Grief and pain are common examples of feelings that are communicated, and often concern for a wounded companion and also communal mourning has been witnessed by ethologists. Hume does not forget to mention negative feelings: "Envy and malice are passions very remarkable in animals. They are perhaps more common than pity; as requiring less effort of thought and imagination."[14]

We can understand this remarkable position Hume takes where the emotions of animals are concerned, if we take into account that his endeavour was to build a moral philosophy on principles as certain and as general as those upon which rests the mechanical system of Newton. The most fundamental "laws" governing behaviour therefore had to be applicable to all men and even to other species. However, there are other reasons.

Hume was an atheist; he considered that the existence of God was a hypothesis that could not be proven, and though he did not say so, for the sceptic that he was it must have been an intimate conviction that this hypothesis was a superfluous one. It is not possible to know how the world came into being, and there are no sound scientific arguments for the belief that it was created following a transcendent rational principle. He rejected the rationalist theory of innate ideas, in keeping with what his predecessor Locke had said about them. In the Christian tradition there is a good explanation for their existence: God has created us in his image and bestowed us with a set of general concepts tailored to our rational needs. Hume, however, considered the Bible as a storybook that fuelled superstition. Without belief in the divine origin of our concepts there is no explanation for the pre-existence in our mind of such entities. It would be unacceptable that they would come out of the blue, but in fact they are produced by our brains. The theory of innate ideas implies the existence of an ontological, religiously founded distinction between animals and men. Its rejection clears the path for a naturalistic approach of the animal species and of our own species as well.

Doubts about the "self" or ego and the immortal soul

Other prejudices obviously favouring the belief of the discontinuity between animals and men are the idea that human beings dispose of a free will and that they possess an immaterial soul, a self or ego. These are other credos that Hume attacks. The novelty of his anthropological views is not due to the fact it offers us a set of new uniquely human characteristics, but rather that he gets rid of a number of *a prioris* that are a hindrance for an empirical approach. He consistently uses Ockham's razor, rejects the existence of essences and of substances other than concrete individual things. He discards the idea that the general has independent reality, and claims that all that exists is particular. Most important is that he adopts Ockham's position concerning the soul, namely that its existence is not given a priori to our consciousness, but can be known only by an act of intuition, in the same way in which we experience what exists in the external world. Hume considers the discussions about the materiality or non-materiality of the soul as irrelevant metaphysical quibbles.[15] Indeed, how can we decide whether something we cannot perceive is material or not? The mind is simply a bundle of perceptions, feelings, images, but many philosophers believe there must be a substance in which they inhere. Hume's reasoning is that knowledge of the soul, as a substance in which perceptions reside, must itself be derived from perceptions:

> We have no perfect idea of anything but of a perception. A substance is entirely different from a perception. We have therefore no idea of a substance. Inhesion in something is supposed to be requisite to support the existence of a perception. We have therefore no idea of inhesion. What possibility then of answering that question, *whether perceptions inhere in a material or immaterial substance*, when we do not so much as understand the meaning of the question?[16]

Hume had to account for the relation between matter and thought. We can suppose that there is no connection between them, that matter cannot cause thought, the former being extended, the latter not. In that case, in order to explain changes in reality, we have to introduce a Deity with an absolute power, being the cause of everything he wills, in the material world as well as in the realm of the spiritual. If God is the cause of all events, he is also the cause of Good *and* Evil, and has willed both, which is a blasphemy. The other solution is to accept that matter and motion can cause thought, and this seems to be confirmed by experience, "since every one may perceive that the different dispositions of his body change his thought and sentiment."[17] He cannot explain this relation between matter and mind in a satisfying manner, but taking refuge in his sceptical theory he declares relations between cause and effect can never be demonstrated and understood, not even where both are material. The Cartesian theory that the mind is unrelated to the body precludes an empirical theory of cognitive processes, while Hume's approach points in its direction.

Concerning the immortality of the soul he is cautious and merely says that we cannot reach a conclusion a priori:

> Any object may be imagined to become entirely inactive, or to be annihilated in a moment; and it is an evident principle, *that whatever we can imagine is possible*. Now this is no more true of matter than of spirit; of an extended compound substance, than of a simple and un-extended one. In both cases the metaphysical arguments for the immortality of the soul are equally inconclusive; and in both cases the moral arguments and those derived from the analogy of nature are equally strong and convincing. If my philosophy therefore makes no addition to the arguments for religion, I have at least the satisfaction to think it takes nothing from them, but that everything remains precisely as before.[18]

Those who believe the soul is an un-extended, indivisible, spiritual substance are prone to identify it with an ego or self, but this is a fiction. Referring to the Cartesians, he says:

> imagine we are every moment intimately conscious of what we call our *self*; that we feel its existence and its continuance in existence; and are certain, beyond the evidence of a demonstration, both of its perfect identity and simplicity. The strongest sensation, the most violent passion, say they, instead of distracting us from this view, only fix it the more intensely, and make us consider their influence on *self* either by their pain or by pleasure. (…)[19]

In fact, however, our self is simply the flowing stream of our changing experiences. It is by feigning the continued existence of the successive impressions, by filling in the gaps that we construct the notion of a soul, a self, a substance. By confounding identity with relation, we introduce something unknown and mysterious connecting the parts.[20] As we know, Hume recognises no metaphysical entities, and in consequence he treats the problem of personal identity empirically.

J.R. Weinberg has shown that in empiricism in general and in Hume's philosophy in particular, much of the scholastic tradition is still present and we can see this from the way many problems are formulated.[21] The main principles and theses of Ockham's nominalism can easily be recognised. On countless occasions Hume bases his argumentation on the principle of *the ontological and epistemological primacy of the individual.* Where our mind is concerned, he stresses that like in other matters we have the natural propensity to consider what shows likeness, be it over time, or situated in space or both, to be identical. Diversity is replaced by identity and this is what induces us again and again into error, also where the self, the ego is treated. Of course human beings are numerically one throughout their lives, but they are not the same all the time, they undergo transformations. We are living beings and though we change continuously, we show a unity over time by the "sympathy" of our parts, which all serve a common end:

> An oak that grows from a small plant to a large tree is still the same oak, though there be no particle of matter or figure of its parts the same. An infant becomes a man, and is sometimes fat sometimes lean, without any change in his identity.[22]

Even changeable objects like a river are considered the same, though it cannot but be in motion and renewing its parts, and though in less than twenty-four hours it is completely altered. The same method of reasoning that we apply to plants, animals, ships, houses, rivers, vegetable and animal bodies must be applied to personal identity. If we say of a person that he is the same person at different moments, we mean that he has the same personality, the same kind of thoughts. This identification is based upon resemblance and causation, rather than, as is the case where material phenomena are concerned, upon contiguity. It is our memory, which discovers the resemblance, between our thoughts, feelings, experiences and their patterns. In fact, our memory produces the resemblance by establishing the relations and from there we conclude to identity. Moreover we imagine the causal link between our thoughts and feelings, how they mutually produce, destroy, influence and modify each other.[23] Hume compares our mind and its components to a republic, in which the members are held together by the same "ties of government and subordination." It may change its members, its laws and constitution and still remain the same unity. Nevertheless, the question whether such a thing as personal identity exists independently, or that it is rather produced than discovered, can not easily be answered. The question is simply to be regarded more as a grammatical than as a philosophical problem.

Free will denied

If there is no such thing as a self, an ego that is given *a priori*, the theory of the free will is jeopardised. Our decisions are taken by the empirical persons we are, not by a fictitious abstract entity, which implies that our physical qualities and our character, which has developed over the years, determine us. He circumscribes "will" as follows:

> I desire, it may be observed, that by the will, I mean nothing but *the internal impression we feel, and are conscious of, when we knowingly give rise to any new motion of our body, or new perception of our mind.*[24]

His reasoning is simple: he distinguishes between indifference and spontaneity, and accuses those who posit the liberty of our will of confounding both. With indifference he means acting without motive, i.e. without cause. Our actions certainly are rarely random, though they can be spontaneous, i.e., unreflective, arising from natural motives. André Gide imagined an "indifferent action," an *acte gratuit*, in which case no moral judgement, reward or punishment would be possible: a man kills a person he does not know and has chosen his victim at random, simply to prove that we can act freely, i.e. without motives. The murderer can in that case not be held responsible, because, Gide says, "the motive is the precondition that is necessary to seize the criminal." We commonly rely upon our experience of the past behaviour of other persons in order to predict their future behaviour, considering what we know from their character, the circumstances and motives. The latter three are considered the cause of their conduct. A person whose behaviour is totally unpredictable is considered a madman. Therefore, those who believe that human beings possess a free will, can only justify this belief by admitting that their behaviour is uncaused, which is not only in contradiction with our common experience, but absurd. Of course we have a false sensation, or experience even, of liberty, "of indifference": "We feel that our actions are subject to our will on most occasions, and we feel that the will is subject to nothing; …."[25] Religion can not be used as an argument against the denial of the existence of a free will: the consequences for religion of a philosophical theory are irrelevant for that theory. That our actions are determined is a thesis founded on our experience of the constant union and conjunction of like actions with like motives and circumstances.[26] If we do deny this, we are obliged to admit that, because our deeds are "temporary and perishing" and have no cause in our empirical selves in the sense explained above, after an evil act we are as pure and innocent as we ever were, because nothing permits to infer from the past that we are going to misbehave in the future.

Hume's moral philosophy is in keeping with his empirical principles: it is the study of the way we judge the character of a person as expressed in the way he behaves. We consider a deed to be good if we feel it as agreeable or useful, bad if we have the feeling that it is unpleasant or harmful. Our feelings are based upon our experience and upon perception. This subjectivism is not solipsism, because we can regulate these feelings upon general rules, and we can try to be consistent with ourselves. Most people seek harmony and collaboration with others. Reason alone is not enough to decide in moral questions; a moral sentiment is needed. This sentiment is love for the person we consider to be good on the ground of our positive feelings, or hatred on the ground of negative feelings concerning bad conduct. From these sentiments follow praise or blame.

David Hume stayed a few years in Paris, where he frequented several literary "Salons." He succeeded in gaining the affection and admiration of the French aristocrats and intellectuals by his simplicity and charm. He was called by his friends *le bon David*, not only because of his quiet amiability, but probably because his slight corpulence gave him an air of *bonhomie*. If proof is needed of his good character it can be found in the benevolence and patience he showed for his friend Rousseau, though this colleague philosopher was notoriously quarrelsome, suspicious and ill-tempered and treated him badly. He certainly was not blind to the fact that aggression, envy and other negative qualities are part of our emotional constitution, but he nevertheless had faith in our general moral capacity. He saw calm passion, dislike of muddle and controversy as sources of morality. Moral virtues and natural abilities, whether of mind or body, were put by him on the same foot. Sympathy was the keyword; to be capable of imagining the feelings and passions of other people enables us to share those emotions, albeit to a lesser degree. If we can step into the shoes of another person we are capable of forming an impartial judgement, and of acting humanely.

Though Hume's philosophy has been of great influence in the domain of epistemology and in the domain of moral philosophy, it never became popular. Like the philosophers of nature in antiquity, like Democritus and Epicurus, he fought superstition; he had no religious faith and offered no comfort to those who demanded of philosophy to be in accordance with their wishful thinking. Instead of lofty speculations about human nature he treated the subject from a common sense perspective and proposed an empirically founded epistemology and moral philosophy.

Notes

1. *Descartes. Oeuvres*, La Pléiade, Gallimard, Paris, 1953, 166–167.
2. *Ibidem.*
3. See J.J. Gibson, *The Ecological Approach to Visual Perception*, Houghton Mifflin Co., Boston, London, 1979, 54, 61–62, and passim.
4. Descartes. *Oeuvres*, 164–166.
5. *Ibidem.*

6. Cf. Donald Davidson, *Tijdschrift voor Filosofie*, 59, nr.2, June 1995 and my reply "Donald Davidson, Descartes and how to deny the evident", *Communication and Cognition*, Vol. 29, nr. 2, (1996), 285–299.
7. Cf. Antonio R. Damasio, *Descartes' Error. Emotion, Reason and the Human Brain*, Papermac, London, 1996.
8. J. Locke, *Essay concerning Human Understanding*, Everyman's Library, Dutton, New York, Dent London, 1967, Vol. I, 116.
9. *Ibidem*, Vol. I, 116–117.
10. D. Hume, *A Treatise of Human Nature*, Everyman's Library, London, 1964, Vol. I, 173.
11. *Ibidem*, Vol. I, 173–174.
12. *Ibidem*,Vol. II, 111.
13. *Ibidem.*
14. *Ibidem.*
15. *Ibidem*,Vol. I, 220.
16. *Ibidem*,Vol. I, 223.
17. *Ibidem*,Vol. I, 235.
18. *Ibidem*, Vol. I, 237–238.
19. *Ibidem*,Vol. I, 238.
20. *Ibidem*, Vol. II, 241.
21. J. R. Weinberg, *Abstraction, Relation and Induction*, University of Wisconsin Press, Madison & Milwaukee, 1963.
22. Hume, *A Treatise of Human Nature*, Vol. II, 243–244.
23. *Ibidem*,Vol. I, 247.
24. *Ibidem*, Vol. II, 113.
25. *Ibidem*,Vol. II, 121.
26. *Ibidem*,Vol. II, 122.

Chapter 7. Idealism and Marxism

Hegel: man by his thought transforms the material world into spirituality

The Age of Enlightenment continued Renaissance humanism. The confidence in man was almost absolute, the coherence of on the one hand knowledge as the product of rationality and on the other the old framework of Divine Revelation, authority and tradition, was lost. Fritz Schalk has expressed this in his characterisation of that period:

> Reason develops freeing itself from the authority of theology, to build the foundation for the emancipation, the autonomy of the person, who, rebelling against a yoke that pressed him down, arises as a power longing to understand the world and master it. The Enlightenment as a fundamentally unchurched world view tries to establish the method of rationality as the root of all knowledge and this not only in the mathematical and physical sciences, but also as the foundation of general laws in ethics, religion, history and politics.[1]

This had been exactly the aim of Hume and of Kant: the former tried to construct a moral philosophy as objective and lawful as the physics of Newton, and the latter, also an admirer of Newton, tried to find the exact limits of human knowledge and the innate universal categories on which it was founded. Neither of them was on good terms with the religious authorities. Hume was known to be an atheist; Kant received from the King of Prussia, though recognising that he was a great philosopher, the advice to abstain from commenting upon religious matters. Moreover, both philosophers believed there was no place for metaphysics in a scientific philosophy or in science.

The confidence in mankind and in rationality, the belief that humanity steadily progresses, is also present in the work of Hegel. He wants to introduce a radically new philosophy that he considers to be the accomplishment and *the end of philosophy* in general. He wants to comprehend the totality of reality, all which in his opinion is of importance in world history: Asian cultures, Greek and European culture. His aim is to grasp the substance of the totality of the profane history of mankind, and then to put it into the frame of the divine process of the production of the Cosmos and of Sacred History. He reproaches Kant that, after investigating the limits of possible knowledge, he has rejected metaphysics as outside its borderlines. He complains that the rational and the emotional have come to be seen as incompatible, that metaphysics and religion are no longer the fertile soil on which science and philosophy grow. Theoretical and practical knowledge prevail, metaphysics declines. It is time to restore metaphysics in its former rights and to construct a broad philosophical synthesis. In his *Wissenschaft der Logik* (Science of Logic), Hegel explicitly declares that for this aim it is necessary to abandon the classical Aristotelian logic, that has been in use for thousands of years, and to replace it with a new dynamic logic in which neither the principle of identity A=A, nor the principle of the excluded third are valid. Parmenides for the first time applied the law of the excluded third in his famous dictum "being is and nothingness is not," but according to Hegel he was wrong: in metaphysics not only being is, but also what is not is.[2]

Of course this has consequences for his vision of man. For him, unlike for the philosophers of the Enlightenment, humanity must not be considered valuable *in se*, but in the light of the totality of Creation. Man is necessary for the completion of God's work: "Wenn Gott nicht wäre, so wäre ich nicht, wenn ich nicht wäre so wäre Gott nicht" (If God did not exist, I would not exist, if I did not exist, God would not exist) he declares, citing Meister Eckhart. His faith is that the world God has created is good, and that humanity will be the instrument of the fulfilment of the promise contained in it. This world view at once meets the desire to *understand* rationally and to *make sense* of reality, which for Hegel is in fact the same thing.

The relation of man and the world

In Western culture man is first and foremost a subject, possessing a subservient body. Descartes proposed a dualistic ontology: the world is composed of two realities, spirituality and materiality. In man the two kinds of reality are reunited: man is composed of a body and a mind. He is a subject knowing and acting upon an external and independent object. Yet, the source of knowledge is not the world, but the mind itself with its inborn concepts and power of reasoning. Sensorial evidence can merely be used to check the theories proposed. Descartes was a man of science, a mathematician and a physicist; his aim was to obtain knowledge that was certain, reasoning that was in accordance with the laws of Aristotelian logic. Scientific theories could be demonstrated; the proofs given could be checked by all rational beings and therefore knowledge was objective. Kant introduced transcendental knowledge, i. e. a critical examination by the mind of its own categories and its own operations. The phenomenal world outside the subject can be known as it appears to us, but not as a "noumenon," i.e. as it is independently from the way it is perceived, conceptualised and represented by us. In consequence the object as it is, "das Ding an sich," escapes the subject, remains external to it. His transcendental critique concerns the universal modalities of human consciousness. Hegel, on the contrary, absorbs the whole world into the mind; instead of existing independently from the subject knowing it, it exists only to extent that it presents itself to the human mind. The mind *contains* the world and *is* this content. However, it is not a passive receptacle, for reality is not given as such to the intellect, but it is *produced* by it. This means that the history of the mind is also the history of the world. Human consciousness must grasp its own truth as it is expressed in the world, in nature, in peoples, in the state, in historical moments. It can understand itself only if it sees the present as the result of the totality of its past, and as a transition towards the future.[3]

In his *Wissenschaft der Logik*, Hegel writes that idealism is the rejection of what is finite, i. e. concrete and particular, as the ultimate reality. This is precisely the sin of nature: it is external to itself, merely given, but unreflected. Even the principles of reality discovered by the first philosophers of nature, Earth, Water, Air, Fire, or the "*apeiron*" (what is indefinite and can give rise to the four elements) are not material as is generally believed, but, being principles, according to Hegel they are spiritual. He declares that the history of philosophy, from its beginning to its end, is in fact the history of idealism.[4] In consequence, in his philosophy *the opposition between subject and object finally disappears*. Man is only human to the extent that he is a consciousness and a self-reflecting consciousness. Thinking the world outside himself, he reconciles the spiritual and the material, the subjective and the objective, the emotional and the rational. Hegel's ambition is to create a philosophical system that is neither subjective nor objective, but a synthesis that transcends both, a vast world view that is religious, metaphysical and scientific at the same time.

Man, however, as said, cannot be understood in isolation from the world; his nature is determined by the place he has in the total reality created by God. The most general feature of reality is that it is in continual change; *all that comes into being must perish*. Whether we consider the history of individuals, of peoples, of nations, or cultures, we see always the same process: they are born, grow, flourish, decay and die, and give rise to ever-renewed forms of themselves. The suffering of humanity is immense and this reflection could certainly lead to pessimism, were it not that in the light of the necessary course of history, pain and sorrow makes sense.

Hegel's starting point is a theological comprehension of history in many, but not all respects kindred to that of Saint Augustine. In the first dialectical moment God before the creation was with himself, contained in himself, One. In the second dialectical moment he poured himself out, creating the world and man, a reality of another order, alien to himself. He incarnated himself by sending his Son to reconcile by his death the world with its Father. The third moment has still to come; it will be the returning of God to himself, the end of history or *eschaton*. In the Holy Spirit, the Creation and God will be united again and the Kingdom of God on Earth realised. This final moment and aim of becoming through time, this absolute end, is the realisation of the Spirit of the World, "der Weltgeist." Man knows only about God, to the extent that God knows about himself in man, and this knowledge is the self-consciousness of God.

Saint Augustine regarded mankind, though originally good, as tainted by the Original Sin. Although Jesus Christ had redeemed our sins, not all was thereby fulfilled; the work was not yet completed, man had still a task. God is Being, and because whatever existed was an emanation of God, as part of Being it had to be good. Also, Evil had to be understood as a gap in reality, a *lack* of being. Without trying to understand God's ultimate reasons, man had the task to repair what is undone, to complete the work of God following his commandments. Worldly history had to be distinguished from sacred history; the driving forces behind the former were human interests, while the latter was the realisation of Divine Providence. Hegel identifies God's freedom with his rationality; whatever happens in the profane world, is justified, because it contributes to the realisation of Reason at the End of Times. Even Evil is justified by contributing to the production of Reason. Instead of dreaming of the original perfect model of reality as being eternal, ahistorical and preceding its realisation, Hegel believes total reality coincides with world history. Containing its own dynamic logic, Creation is not moulded upon eternal ideas; it has not yet come to an end, but it is an ongoing dialectical process evolving ever-new forms. Man is indeed the instrument through which God attains his ultimate goal, but sacral and profane history are not opposed but one, its moments producing an ever wider spiral of *progression.*

As Karl Löwith has shown, the difference between Augustine and Hegel is in fact, that though both consider history from a Christian perspective, the former sees in it the expression of the Absolute Power of God, while the latter sees in it the work of the World Mind, which through the "cunning of reason," using the passions of men, attains its goal. It is not by chance that the result of the great human enterprises is never quite what the actors involved had meant. Thus, says Löwith, Caesar and Napoleon did not know and could not know the consequences of what they achieved, when they founded their empires. They were not aware of the fact that what they planned had to be fulfilled, because it was part of the general aim of the history of the West. All events correspond to the iron logic inherent in reality. It is this ambivalent freedom of the passions that creates a world that is not yet understood as the expression of the underlying teleological principle. Indeed, the seemingly free creations have, of necessity, as a moment in the dialectical progression of humanity, to be accomplished.[5] The plan of the World Mind surpasses all plans on the human scale, and whether individuals rebel unknowingly against it or obey it, they are only its instruments. Thus, looked at with "the eye of reason," history is the expression of rationality, or, what is the same, Divine Providence.[6]

From this point of view World History is also World Tribunal: whatever is accomplished serves a higher and hidden aim, and is *justified* by this fact. Karl Popper in *The Open Society and its Enemies*[7] severely criticises this theological and teleological conception, warning for the danger of totalitarianism it contains. All events must be understood in the light of the Original Myth, the Christian world view, and then they make sense. The individual has in this conception no value in its own right. On the contrary, its value is identical with that of the place it occupies in the dynamic system, contributing to the inexorable progression of history. Indeed, Hegel saw the particular individual with its own interests as the thesis, the community with interests opposed to those of the individual as the antithesis, and both were surpassed and reconciled in the synthesis, the State.

In *Die Vernunft in der Geschichte* (Reason in History), Hegel explains that if we ask what is the ultimate goal of Reason the answer is that it is the same goal as that of the world. Indeed, Nature is in his conception only ideal *because it is such that it can be thought.* In his philosophy the duality between the Mind reflecting upon Nature and reflecting upon itself disappears. However, World History is the main subject of his investigations, and though physical nature plays a role and, when a catastrophe occurs, interferes in the course of World History and can alter it, what is substantial is the Mind and its development. He concludes that therefore his task is not to consider nature *in se*, objectively, but only in as far as its rationality can be understood. Thus, nature too is a system of reason, special of course, idiosyncratic, but always *existing only relative to the mind.*

This is how Hegel characterises man in relation to nature and mind:

> After the creation of nature man comes into being, and he makes out the opposite of nature; he is that being that rises into the second world. We have in our general consciousness two realms, that of nature and that of the mind. The realm of the mind is that what is produced by men. One can make all kinds of representations of the Kingdom of God, but it is always a realm of the mind that is realised in man and must be brought into being by him.
>
> The domain of the mind is the most encompassing; it contains all that has ever interested men and still interests them. Man is active in it; whatever he does, he is a being in which the mind is working. Thus it can be of interest to recognise in the course of history the spiritual nature of its existence, i.e, how the mind is united to nature, or, in other words, human nature. When one talked of human nature, one has imagined something that was constant. The representation of human nature has to apply to all men, from past and present times. This general representation can undergo infinitely many modifications, but in fact all of them are representations of one and the same being under the most different forms. It is the thinking reflection that leaves out the distinctions and retains only what is general, which under all circumstances will show itself as active in the same way and in the same interest. The general type can also be indicated in that what seems to be most alien to it; in the most crooked human form its humanity can still be recognised. The fact that in such a form still a mark of humanity remains, can present a kind of consolation and reconciliation. In the reflection on world history from this point of view, it is stressed that human beings have remained equal to themselves and that under all different conditions their vices and virtues were the same, and one could appropriately say with Solomon: there is nothing new under the sun.[8]

As could be expected, the distance between the anthropology of Hegel and the view upon man of the empiricists is enormous. To the latter a living being is nothing but a concrete individual, a body receiving stimuli from its senses. It is provided with intellectual and emotional capacities in accordance with the richness of the information of the outside world that reaches it. Its bodily condition determines these capacities and in man the quality of his humanity. Hegel, in contrast with this view, can recognise the general essence in the most deformed human creature, because a human being *has* a body, but *is* a soul. For the idealist generality always prevails upon individuality, the spiritual upon its material realisation.

Also, it is not surprising that Hegel rejected the theory, which was popular in his time, that a man's intellectual and moral capacities could be derived from the particular measures of different dimensions of his skull.[9] The body is an instrument for the soul that enables it to become a concrete presence, but the body cannot determine our personality. Yet things are never simple in Hegel's philosophy and all oppositions must be reconciled in the end. The individual is determined by certain external elements, the place where it is born, the climate, etc.; it belongs to an ethnic group, with its own culture; on the other hand the individual is capable of autonomous free action. *In itself* it is an individual body with particular characteristics related to its surroundings, but *for itself* it is consciousness guiding and moulding materiality. It is both indeed, and both are needed in order that it becomes what it is through its actions. The distinction between *an sich*, in itself and *für sich*, for itself, is crucial. The ego never completely coincides with itself, because as soon as the individual expresses and externalises itself in its deeds, it does no longer recognise itself as such in them:

> Thus the deed as finished work has the double opposite meaning, to be either the *internal individuality* and *not* its *expression*, or to be, as something external, a reality *free* of what is internal, something utterly different from it.[10]

The deed as expression of our inner self in its concrete realisation by our hand, our mouth and our voice, by our legs, our gesture or expression of our face, *exists for others and becomes alien.* We can reflect upon our psychical identity, even reflect upon this reflection, and we can use our actions as a creation of a self we are not, a mask. By this self-reflection we internalise again our external self that is constituted by the characteristics that are genetically determined and determined by our education; we internalise the expression of their combination in the manifold of our interventions in reality.[11]

But let us not forget that the individual is only the raw material of history, that it must be reconciled with what is general, not only with the common good of society, but also with the general purpose of history. Individuals are not the ultimate reality, individuation is never completed in Hegel's thought; in fact, matter, not as a principle, but individuated matter, does not really exist. The external world consisting of particular individuals in time and space is evil. A certain manicheism, the spirit being what is good, matter what is bad, is not far away.[12] Hegel considers the empirically given as "gemein," as muddy and murky, imperfect: it is not, as Augustine had said, evil that does not exist, but matter that is non-being.

On freedom

In his *Wissenschaft der Logik* Hegel makes it clear that, though God would be the most worthy subject to begin his exposition with, it will be the concept of Being that he is going to treat of first, because it is the most empty of all concepts, equalled in this respect only by its contrary, Nothingness. In a dialectical movement Being turns into its antithesis, Nothingness, and both are reconciled and surpassed in Becoming. We have seen that history is a theodicy that uses human freedom to attain its goals:

> When the mind aspires to its centre, then it aspires to perfect its freedom; and this aspiration is essential to it. If one says namely, the mind is, this means at first sight that it is something that is finished. But it is something that is active. Activity is its essence; it is its product, and then it is also its beginning and its end. Its freedom does not consist of a resting being, but in a continuous negation of that what threatens to abolish freedom. To produce itself, to make itself to the object of itself, to know about itself, is the affair of the mind; thus it is for itself. Natural things are not for themselves, and therefore they are not free. The mind produces, realises itself following its knowledge of itself; it works in such a way that what it knows of itself is also realised. Thus all that matters amounts to the consciousness of the mind of itself; when the mind knows that it is free this is something totally different from when it does not know it. Because, when it does not know it, then it is a slave and contented with the slavery and does not know that this is not as it should be. It is only the experience of freedom that makes the mind free, though it is in se and for itself always free.[13]

This view has been most influential; Hegel makes *self-consciousness and self-reflection the hallmark of true humanity*. That we are not merely there, but *that we know about ourselves*, that is our liberty. The limit of our freedom are unconscious processes, and amongst them the working of instincts.

All living beings, says Hegel, have instincts, including man. The objects of our instincts are sensorial. By being directed toward these objects that are external to us we become also external to ourselves. Like our observations our instincts are single, particular. If a man was identified with his instinctive life he could not be distinguished from the animals, because in instinctive actions there is no self-consciousness. Human beings, however, are self-conscious and can grasp what is general and ideal. They are thinking beings and as such know about their instincts and therefore can give in to them or refrain from doing so. Because they are not abandoned to their sensorial experiences and their instincts, they cease to be merely natural creatures. They set themselves aims and thereby direct their mind upon what is general instead of what is the particular object of their desires. By representing for themselves their aims and actions and choosing amongst them, human beings determine what they will be. Animals cannot put their reflection between them and their urges; the connection cannot be disrupted, but by pain or fear. The autonomy of man consists of knowing what determines him and of being able to thwart this determination. This is his positive freedom.[14]

Thus, Hegel considers human freedom to be the motor of history, but in contrast with the normal conception of historical processes, at the same time he conceives them as teleological; their outcome is not the result of chance, but is necessary. For most of us history is not a mechanical and therefore foreseeable process, but each individual combined with the circumstances of its existence, constitutes a causal chain that interacts with innumerable other similar causal chains. This is what makes history a chance process, and though it shows regularities, it cannot be predicted. From Hegel's teleological point of view, man is not the play-ball of history as such, but, as we saw, of the "cunning of reason," which turns, without the actors knowing it, all events into contributions to the realisation of the ultimate aim.

In as far as nature is matter, it is death, but in fact it contains (potentially) life, i.e. the germ of spirituality that slowly and progressively grows and produces a diversity of forms and eventually man who produces mind:

> Nature is *in itself* a living totality; the movement through its gradations is essentially that the idea *posits* itself as that what it is; or, what comes to the same, that from its immediacy, and externality, which is *death*, nature turns *into itself*, to be thereafter living, but furthermore also that it lifts its being by which it is life, and raises itself to the existence of the mind, which is the truth of Nature and the true reality of the Idea.[15]

The origin of species

It is interesting that Hegel's historicism did not lead him to consider the origin of species as a historical evolutionary process. This is because history is the expression of the Mind at work, while Nature is the contrary of spirituality. If it contains spirituality it is only potentially. Indeed, nature is non-idea, it is there, outside; it can be defined as "*externality.*" Nature is such that we can think it, but because this spiritual aspect of Nature cannot reflect upon itself Nature is not free, but at once necessity and chance. The peculiarity of Nature is that it is on the one hand given and is as it is, determined by the laws that apply to it, and that on the other hand it is the result of chance, but it is never its own idea. This is its shortcoming: its being a non-idea, and thereby a non-being. Nature is the betrayal of the Idea. Only in our primitive consciousness, our sensual awareness of the world, Nature is the first, immediate being. Though a divine idea has preceded the natural world, it is far more difficult to recognise the Being of God in natural phenomena, than in our poorest representations, in our most imperfect fantasies.[16] In Nature there is also a contradiction; it is divine in the idea, but its being does not correspond to its concept, because it is not ideal, but material, particular and concrete. It is the idea perverted of itself. Not being self-conscious it is a mere datum without conceptual dimension and therefore non-being. The highest form of nature is life, but even this form remains abandoned to the non-intelligence of its externality. Only a spiritual expression contains the moment of a free general relation to itself.[17] The path of the celestial bodies, the growth of plants, the life of animals are governed by both chance and necessity, whereas human history is the result of human reflection and decision and though often seeming irrational it is ultimately governed by Providence, and therefore superior. History leads towards a goal that will of necessity be realised at the end of times.

Hegel rejected emphatically the views of the ancient philosophers of nature, who explained the origin of species naturalistically.

> It has been an unsuitable representation of old and also of new philosophy of nature to regard the further elaboration and the transition of a form and sphere of nature in a higher one as a production in the real external world, which in order to make it clearer has been relegated to the darkness of the past. It is a proper characteristic of nature to let, in its externality, fall the distinctions apart, and to let them appear as indifferent existences; the dialectical concept, that conducts the steps, is the internal aspect of it. Such nebulous, fundamentally sensual representations, like more especially the so-called arising e.g. of the plants and the animals from water and then the arising of the more developed animal organisations from lower ones, etc., must be avoided by the thinking reflection.[18]

Hegel admits that it cannot be denied that even Nature contains its own idea, but only in an abstract way, and not in its concrete, individual realisations existing in their variety and oppositions without further connection. Necessity is to be found in the functionality of forms, chance in the lack of regularity and determination of its creations. The contradiction of these opposite poles remains unresolved in the concrete world. There is no logic in it, while, in the realm of the spiritual, dialectical reason progresses by transcending thesis and antithesis in a synthesis, in which both previous moments are still present but surpassed in something new. Nature is incapable of building forms of life using the abstract idea inherent in it as a guideline. It is full of exceptions, derogations, and failures. We cannot derive species, classes and orders from the empirical data, which proves the autonomy and dignity of the conceptual order.[19]

Hegel criticised Newton's mechanical system, because it did not correspond to his idealistic model of physical reality. For the same reason he missed the chance to prepare and facilitate the biological theory of the future: evolutionism. Moreover, his anthropology can be considered a restatement of the traditional Western conception of man, along the lines of his dialectical idealism. Man is still characterised as anti-nature, as a rational being that is self-conscious, as free and historical in essence. New is that subject and object are reconciled: the subject contains its own object, the objects exist through the subject. Also new is that the relation of man and God is no longer asymmetrical, God being the sole source of Being. Hegel has absorbed in his philosophy many ideas of the Enlightenment and accords to man a prerogative of God: mankind becomes a second source of Being.

It is easy to recognise traits of pantheism; God and the world, conceived as a continuous becoming, coincide. Hegel interprets the belief in a God to be the expression of what he calls the "unhappy consciousness": men considering the world and themselves as unsatisfactory cannot understand themselves but as the unworthy creatures of a perfect being, a transcendent God. This is only a stage in the long way man has to go before he becomes conscious of his autonomy. God, infinite Being, is not present *before*, but *in* the finite world; Reason does not precede the world, but is produced by it. God is God in as far as he has knowledge of himself in the spiritual being that is man. The historical dimension of an individual is not to be understood in the ordinary sense that he can change things, achieve goals according to his own plans, but in the sense that he is part of the vast dialectical and teleological process that is a continuous *progress*. History for Hegel is not rectilinear, but it has the form of a spiral steadily becoming wider and leading to more and more spirituality.

The end of the alliance between philosophy and science in continental philosophy

Hegel's influence has been immense in his own time and long thereafter. His followers can be divided into the "right-wing Hegelians," who were proponents of a conservative interpretation of the Master, and "left-wing Hegelians," who were inspired by him, but who also opposed him. Amongst the latter can be cited Bruno Bauer, Max Stirner, Ludwich Feuerbach and Karl Marx. By the end of the nineteenth century his popularity had greatly diminished in Germany, his philosophy was often misunderstood and ridiculed. In Great Britain, however, the "bulwark of empiricism," Hegelianism was represented in the second half of the nineteenth century and the beginning of the twentieth century by men like Bradley and Mac Taggart. In the twentieth century a revival took place in many countries. German idealism was continued in the work of E. Husserl, whose phenomenology gave rise to existentialism. The influence of Hegel on existentialism, especially on the work of J.P. Sartre, is well-known. Moreover, for Marxist intellectuals in Europe the study of this source of the doctrine was a necessity.

After Hegel philosophy is finally divorced from science on the continent. Not that it was Hegel's purpose to contribute to this evolution; his intention was just the reverse. Nevertheless, the progress in many domains of science was accelerated in the nineteenth century, and not only the gap between religion and science, but also between metaphysics and science became ever wider. We can consider that Hegel's system was the last major attempt to unify all three. He could not reverse the process that was ongoing: science developed its own methodology. Its logic was Aristotelian not dialectical, and metaphysical and religious considerations could no longer be taken into account in the construction of scientific theories. Nevertheless, a century later Husserl still dreamt of a "first philosophy" that would be the foundation of all the sciences, but this remained a dream. His enterprise was completely ignored by the natural sciences. It seems that for many philosophers it was difficult to understand that the battle was lost. However, the evolution of science could not be reversed; it had to gain its autonomy over the centuries, and one of the conditions was that it had to delimit its domain very carefully, and to ban the subjective point of view. Science became not only autonomous but also self-sufficient. The champions of metaphysics in continental philosophy took a defensive position, or were even hostile to it.

The divorce between science on the one hand, metaphysics and religion on the other hand, led to the curious phenomenon that the West developed two different and incompatible pictures of man. Most continental philosophers continued the millenary metaphysical tradition and the further elaboration of its conception of man. Only a few philosophers offered us an alternative: Nietzsche and Schopenhauer.

Marx: man humanises nature by his production

Hegel and Marx

The source of the philosophy and anthropology of Marx is in the first place Hegel and in the second place Feuerbach; from them he inherited the humanism of the Enlightenment, the conviction that from all times *man is what is of interest for man*. This means for him, like for Hegel, that the rest of reality must and can only be considered to the extent that it is related to man. Marx was convinced that the great step forward he took was to have rendered to human beings their concrete existence in two respects: in the first place by examining their ties with nature to the extent that they continually transform it, and in the second place and most important by restoring to them their historical and socio-economic dimension. He also believed that Hegel's philosophy remained abstract, because it started from a religious and metaphysical world view, deriving from it the nature of man. He, on the contrary, derived his conception of man from the study of historical documents, but also from the study of concrete economical and social situations that were at hand, right before his nose, revealing to him the human condition. This is the meaning of his famous dictum that Hegel had turned dialectics upside down:

> The mystification that dialectics suffers in the hands of Hegel, in no way prevented that he has presented for the first time consciously and globally the general forms of movement. It is upside down and has to be turned right again with a view to discovering the rational core under the mystic covering.[20]

Hegel's religious and metaphysical theory precedes the praxis, while in Marx's own philosophy the praxis determines the theory. Marx rejected idealism and replaced it, not by the materialism of his time that he scorned for being gross and vulgar, but by a new *dialectical* materialism.

He claimed that his method was truly scientific; it was no longer subjective, but based upon material facts. Yet, this suggests immediately the question whether dialectical reasoning can in principle be scientific, considering that its logic is utterly different from the Aristotelian logic used in science, as Hegel had explicitly stated. Marx did not pose the problem in these terms. On the contrary, starting without hesitation from the dialectical method, he was certain to turn it into a scientific tool, anchoring it, in contrast to Hegel, in concrete reality. His reasoning was dialectical because reality was, not the other way round. Indeed, the oppositions were not those between concepts, the true nature of reality being ideal, but oppositions existed independently from the human mind, in the concrete world. Genuine contrasting poles could be observed in man, nature and society alike, creating accumulated tensions, which eventually had to be resolved. Reality moves on from dialectical moment to dialectical moment, generating ever-new moments in which the oppositions are included, transcended and reconciled. Like Hegel, Marx had an encyclopaedic mind and wanted to create a new science at the service of man, encompassing all knowledge in different domains, which would be the raw material to be processed by dialectical materialism. Thus he hoped to be able to predict with certainty the outcome of historical processes, to prove his theses and, once this work done, to derive his philosophy and his anthropology from them.

Science must be historical science

Like for many authors different periods can be distinguished in his work and this must be taken into account. In *Die Deutsche Ideologie* (German Ideology), one of his *Frühschriften* (Early Writings), the basic ideas of his anthropology can be found, whereas from the *Manifest der Kommunistischen Partei* (1848) on, a few years later published as *Communist Manifesto*, he became more involved with political and economical theories, instead of with a prophetic interpretation of history. In his early work, he explains that the all-encompassing science he wants to develop is centred on the history of man, not on that of nature:

> We know only one science, the science of history. History can be considered from two sides and then divided into the history of nature and the history of men. However, both sides are not to be separated from time; as long as men exist, they mutually determine each other. The history of nature, the so-called natural sciences, are here not our concern; but we will have to go into the matter of the history of man, as almost the total ideology is reduced to a biased decomposition of this history or to a complete abstraction of it.
>
> The presuppositions from which we start are not indifferent, they are not dogmas; they are real presuppositions, of which one can only make abstraction in the imagination. The individuals are the real individuals, their activity and their material living conditions, as well those that are already given as those that they have produced by their own activity. Therefore, these presuppositions can be established purely empirically.[21]

Though mankind certainly must be studied as a species, i.e. a collection of concrete individuals, he does not want to study their biology, their bodily condition, nor their environmental conditions, he does not use these elements to explain the nature of man, but presupposes them:

> Naturally the first presupposition of all human history is the existence of living human individuals. The first fact we have to witness therefore is the bodily organisation of these individuals and their relation to the rest of nature that is thereby already given. We can here of course neither enter into the matter of the physical qualities of men themselves nor into the matter of the natural situations men encounter, the orohydrographic, climatic and other conditions. All historiography must presuppose these natural basic facts and their modification in the course of history by human activity.[22]

In spite of the fact that Marx recognises that men are embedded in nature, his anthropology is clearly anthropocentric. As said, like for Hegel what is of interest for men are men.

When reflecting on the originality of Marx' contribution to anthropology, this could be a very good synthesis: instead of treating of the abstraction "man," he wants to treat of individuals whose reality is that they are embedded in a socio-economic system that determines their living conditions and therefore themselves. For Hegel what matters is not a concrete individual, but man in his generality, the maker of reality, as it were by secreting concepts entering dialectical relations that produce ever more concepts, a process that will finally realise the World-Mind. Marx is interested in individuals working under specific economic and social conditions, earning their living, from which depends what foods and clothes they can buy, and whether their children can see a doctor if necessary and whether they can go to school. What these men and women of the working class produce are not ideas, but goods. What they are depends in the first place on what they do, and only in a second way on what they think. And what they do depends on their social position and the socio-economic situation. Each man is not only what he actually is, but also his possibilities, what he can achieve for himself and his family within the limits of his social condition. These are not loose remarks about human beings and the role they play in society, but this is the basis for a definition of humanity, as will become clear from the next section.

Man defined as transforming nature by his labour

Now Marx comes to the crucial question: what is peculiar about man and distinguishes him from the animals? He is very clear about this:

> One can distinguish men from animals by their consciousness, by their religion, by whatever one wants. They themselves start to distinguish themselves from animals, as soon as they begin to *produce*, a step that is conditioned by their bodily organisation. If men produce their necessities of life, then they produce indirectly their material life itself. The manner, in which men produce their necessities of life, depends in the first place on the kind of natural goods they find and want to reprocess.
>
> This manner of production is not to be considered only from the side of the reproduction of the physical existence of the individuals. It is rather a definite nature of the activity of these individuals, a specific way to express their life, a lifestyle that is specific to them. How the individuals express their lives, thus they are. What they are coincides with their production, as well with what they produce, as with how they produce it. What the individuals are, therefore, depends from the material conditions of their production.[23]

The relation is symmetrical; human beings are defined by the way they transform nature, but they are also defined by the opportunities nature offers them. The production of goods is only possible when the population of individuals grows and through the commerce individuals have with one another, commerce that is in its turn conditioned by the kind of production in question. Of course men do not only make material goods, but also ideas, which have no independent existence, but which, as the result of human activity, are limited by the social and economical situation of the men who produce them. Crucial for these broader social relations is the division of labour that takes place in different steps. First it exists in its most primitive form by the different roles of men and women in the family, next when individuals collaborate as members of a society and develop specialised ways of producing different kinds of goods, and finally further division takes place when social relations show more and more complicated forms. From the moment the division between spiritual labour and material labour is a fact, consciousness can imagine itself to be different from the existing *praxis*: from then on it is possible for men to imagine that they are something that is different from what they do, but this is an illusion. They think they can create something that is only a representation, without being aware that what they have created is something that is as real as all the other things they produce. That is the moment that in their consciousness they are capable to emancipate themselves from their primitive human condition and to build theories, philosophical systems, ethics, and religion.[24]

For most historians historical times begin with written language, for Marx, however, they begin when man *produces* the goods he needs for survival instead of hunting and gathering food and other things as he finds them. As a counterpart to Hegel's idea that man can transform matter into spirituality by thinking it, Marx believes that it is the task of man to humanise nature by transforming it by his labour. On the one hand Nature is not completely material, on the other hand consciousness is not pure, but tainted with materiality. This materiality of the mental can be witnessed in its expression in language. "Pure" mind does not exist; it is tainted with materiality, because it can only be known if it is expressed by the vibration of air, by sounds. In short, mind exists in language, i.e., practically speaking, it exists as it exists for others. Marx gives credence to the common theory that language is as old as consciousness; both have developed, because men need to relate to other people, to communicate. The animal has, in his opinion, no relationships to other animals to be compared to the social relationships between men. Because it cannot grasp them, they do not exist for it.[25]

In *Das Kapital* (The Capital), he explains clearly the difference he sees between the work of men and that of animals:

> Labour is in the first place a process between man and nature, a process in which man by his own activity realises, regulates and controls his metabolic exchange of matter with nature. He opposes himself to the raw material as being himself a force of nature. He begins to move the natural forces that belong to his corporeality, his arms, legs, head and hand, in order to take in possession raw matter under a form that is useful for his survival. When by this movement he acts on nature outside himself and changes it, he does so dialectically, changing at the same time his own nature. He develops the potencies that lie dormant in him and subjects the interplay of its forces to his own project. Here we are not in the presence of some sort of primitive, animal, instinctive forms of work. The labourer, as the seller of his own working power on the labour market, has against the background of primitive times escaped the situation in which human labour had not yet shed its first instinctive form. Here we assume labour under a form that belongs only *to man*. A spider executes operations that resemble those of the weaver, and a bee by the construction of its cells of wax puts to shame many a human architect. What, however, distinguishes from the beginning the worst architect from the best bee, is that the architect has built the cell in his head, before he builds it in wax. At the end of

the working process a result appears that in the very beginning was already present in the *mental representation of the worker*, and therefore was present *ideally*. It is not the case that he only *causes* a change of form of something that is natural; he *realises* in what is natural at the same time *his aim*, because he *knows* the laws determining his ways of acting to which he must subordinate his will. And this subordination is not a singular act. Next to the effort delivered by the organs that work is the *purposeful* will present, which is expressed as the *attention* necessary during the total time the work takes, ...

The singular moments of the working process are the *purposeful activity* or the *labour itself*, its *object* and its *means*.[26]

According to Marx, for man to exploit nature is in accordance with the general order of things. Although in the nineteenth century the situation was already deteriorating in many respects,[27] there is no sign of ecological consciousness. This is in accordance with the anthropocentric view that the only thing that matters to man is man and that the natural resources are there for us to exploit in unlimited quantities. He declares:

The earth (economically including also water), which originally provides man with victuals, necessities of life ready for use, without his intervention is found as the general object of human labour. All those things that the work of men breaks loose from their direct relation to the totality of the earth are naturally existing objects to work upon. Thus the fish that is separated from water, its vital element, and caught, wood that is cut in the jungle, mineral that is cut loose from its vein....[28]

The alienation of man

As we have seen, unlike Hegel, who believed the fate of an individual is determined by the historical processes of which it is the play-ball, Marx's approach is less abstract; his thesis is that the fate of an individual in the first place depends on the social class to which he or she belongs. In a developed society the individual is alienated, i.e., no longer able to be in harmony with its own nature. At the origin of human alienation is the division of labour: once this division was accomplished in a given society its members could no longer freely decide what they would become, what they would do for a living. Originally, it was nature that imposed what kind of work men had to do in order to survive, but as society became more complex it imposed, depending on what position an individual occupied on the social ladder, what work he or she would be obliged to take on.[29] Living in an industrialised and capitalist society, it was clear as crystal that most men could neither decide for themselves nor by a common consent what and how they were going to produce goods. They became dependent on external powers in society, which they did not understand and could not alter. They no longer chose and imagined what they were going to make, they no longer possessed what they made, but their creations began to live a life of their own, and they got out of control, materially and mentally.[30] For Marx this alienation will be the motor of history, which he conceives as the struggle of those who do not possess the means of production and those who do, i.e. the struggle of classes. History is on the one hand deterministic in a teleological sense, because it is certain that the exploitation of men by men will, when reaching extreme proportions, lead *of necessity* to a world revolution. At the end of times a society without classes will be realised. It will not be the Kingdom of God on Earth, but the Kingdom of Man on Earth. Then men will possess in common with all other men the richness of the earth, all the goods they desire. Freed, not of his material needs, but of their tyranny, a man will be no longer alienated and a slave, but a free human being choosing what he will become. In the struggle, millions are crushed like ants under the heavy boot of history, but this is of no importance if it can further the holy cause. On the other hand the individual, becoming conscious of the process he is *volens nolens* part of, is indispensable for the realisation of the ultimate goal. This is the new eschatology Marx derives from his analysis.

Nature dialectically related to Man

Marx had equally new ideas about Nature. He recognised that he was indebted to Feuerbach, in this sense that the latter had shattered the system of Hegel and replaced it by a theory restoring reality to what is individual and concrete, drawing together again the subject with the object, mind with matter and man with nature. Yet, he criticised Feuerbach's conception of matter as such, unrelated to man, but also his conception of the reality of what a human being is. Indeed, Feuerbach's viewpoint was in his opinion anti-dialectal and ahistorical and therefore remained abstract. Just like man had to be considered in his relation with nature, nature had to be considered in its relation with man, as such it is a non-existent:

> This activity, this continuous sensual working and creating, this production, is so much the fundament of the total empirical world as it now exists, that, if it was interrupted for only a year, Feuerbach would witness an unbelievable change not only in the material world, but that he would also no longer recognise the world of men and his own way to look at it, and even his own existence. Anyway, in this the fact that external nature is prior remains the same, and moreover all this does not apply to the first men, procreated by *generatio aequivoca*, but this distinction we introduce here makes only sense if man is conceived as being different from the animals. For that matter, this nature in which Feuerbach lives and that precedes human history, a kind of nature that, except on a few Australian coral isles, nowhere exists any longer, in consequence also for Feuerbach does no longer exist.[31]

It is the essence of man to have, while being part of nature, a dialectical relation with it, to transform it and to humanise it and it is wrong to understand man either as opposed to nature or as its mere product.

Religion

Although Marx was convinced that his view of the world and of man followed from truths he had previously empirically and rationally established, many critical commentators have doubted this. They have suggested that he merely transformed the Christian eschatological theory of Hegel into a materialistic, but at the same time a dialectical, and therefore teleological, account of the history of man and its ultimate purpose. Anyway, it must be said that his world vision was at the same time rational and more than rational, in fact a confession of faith. He constructed a vast well-documented theory that also purported to give sense to human existence; he was a scholar *and* a moralist *and* an activist. In other words he was objective and at the same time he was not, because he mingled "is" and "ought," and even believed that what the facts were and how he thought they should be could be proven to coincide in the end.

Of course, if it is the upper-class in a society that determines its ideology in function of its interests—as expressed in the succinct formulation by Marx: "they say God, but mean cotton"—and if it is the task of the Marxist to unmask this ideology and show it for what it is by a thorough critical examination, the question must be asked "if the truth is not the truth of the upper-class, is it then the truth of the working-class?" The answer for Marx is that the proletariat has nothing to lose, neither power nor possessions to defend, and therefore its point of view is not biased by interests. If the method of dialectical materialism is scrupulously followed, the point of view of the worker is the general point of view, the view of all men, and therefore superior to that of a wealthy minority.[32]

We have seen that in Hegel's philosophy man exists because God exists, and that if man did not exist, God would not exist. Marx followed Feuerbach, who reduced God to an enlarged projection into a transcendent reality of the image human beings have of themselves and also of their characteristics, Reason, Love and Willpower. Marx adopted this idea, but he refused to see in the irrational longing of man for a better world the essence of humanity. Of course religion is created by men seeking solace for their suffering in the promise of a better life after death, but the source for religious faith is not to be found in the essence of man, but in his *alienated conditions*. Because God is a fiction, in Marx's anthropology the original opposition shifts from that between man and God, to that between man and nature. What is unique in man is that he *is* at once nature and is *opposed* to nature, and that he resolves this opposition by transforming, in a long historical process, nature into culture. Though to this opposition different solutions can be found, it can never totally disappear; more even, being the foundation of all other oppositions it cannot be resolved.

Historicity

A consensus seems to exist among specialists that the concept of "alienation" was the motor of historical change in Hegel's philosophy, and, as we have seen, it is also basic in Marx's account of human reality. He believed that the first men had not yet discovered their true humanity. They were ahistorical like the animals, because they were not opposed to nature, did not dominate it, but were part of it and lived of whatever nature offered them. As hunters-gatherers, they lived in a condition that may seem to some comparable to that of Adam and Eve in Paradise, but to Marx was pre-human. It is only when man becomes sedentary, grows wheat and breeds cattle, that he becomes truly human. The first historical event is the production by man of his means of subsistence. Therefore, as we already mentioned, prehistory does not end with the invention of writing, but when man by his labour subjects nature to his needs. This is what German philosophers, being still idealists instead of materialists, could not understand. Marx comments ironically:

> With the Germans "who do not start from presuppositions" we must begin with establishing the first precondition of all human existence, and therefore also of all history, namely the precondition that people must be able to live in order to be able to "make history." To life, however, belongs before all else food and drink, housing, clothing and some other things.... Here is shown at the same time whose brainchild the great historical wisdom of the Germans is, for whom, where they miss positive material and neither theological, nor political, nor literary nonsense is traded, no history whatsoever, but prehistory takes place, and this without explaining to us in the meantime how one can pass from this nonsense of "prehistory" to genuine history.[33]

It is not surprising that Marxists cannot accept the idea that men who cannot read or write, but live in societies with extremely complex structures, and have developed sophisticated cultural systems, nevertheless are conservative to such a degree that when forced to adapt to new circumstances, they only transform their society and culture, but no sudden fundamental changes occur. Claude Lévi-Strauss was severely criticised by the Marxist intellectuals of his own time for having suggested that to be historical was not an essential property of mankind. To him, who admired the great varieties of cultures he had encountered, each culture was essentially a way to cope with the geographical, climatic situation, with the economical means available in the natural environment, a way to create stable social systems that enabled men to survive. The primitive men Lévi-Strauss studied, though not necessarily living in paradise, were not alienated in the Marxist sense, and therefore alienation could not cause their socio-economic systems to be swept away and to be replaced by a new system that would be the next step in the historical progression of mankind.

Is Marx's concept of man revolutionary? In fact it is in some respects. Marx declares explicitly that *man is what and how he produces* and thereby his *materiality* is restored to him, not as an original sin, but as an essential part of his humanity. However, Marx does not yearn for the returning of the primitive condition of men, because, though forever embedded in nature, all men at the same time are bound to leave their aboriginal state behind them, to domesticate their own nature and external nature, to use them for a definite purpose. Men cannot be defined by their place in the static cultural and social system they are part of, except when they still belong to an unchanging primitive society and are still in a kind of pre-human stage. They must be defined as *historical* beings, whose individual lives make sense in the perspective of their role in the changing of society and the perspective of the future of humanity. They have a *task*; they must contribute to *the progress of mankind* that is inherent in the general course of history. From the essential thesis and antithesis, from materiality and spirituality, present in man at the same time, follows the replacement of the direct use of what nature has to offer by the manufacturing of goods by means of production in which both principles are reconciled. When men produce more than they need to survive, the division of labour becomes possible, and they can sell their labour. The more complicated social relations between those who buy work and those who sell it are what causes the alienation of both buyer and seller. The true nature of man will be revealed in a society without classes, where the interest of the individual is in accordance with the general interest. Then man overcomes his alienation, his dependency upon what is external to him and he can benefit freely from what society produces, choose the kind of labour he wants to perform for the realisation of aims he has under control, and finally enjoy leisure. Like the heathen Aristotle, Marx believes that man is not condemned by God to work in the sweat of his face; only, to him leisure must not be the privilege of the upper-class in a society economically based upon the toil of slaves, but the right of all men.

After his death Marx became more famous than he had been during his lifetime from the moment political movements claimed his ideas. His followers put his theories or their interpretation of them into practice. The success of the Bolsheviks in 1917 turned Marxism into the ruling ideology in Russia. Under the form of Marxism-Leninism it became the orthodox ideology in the Soviet Union and in China. What concerns us is his anthropology, not his political and economical theories, but as the lives of millions of peoples was affected by the application of his ideas, the question cannot be avoided whether his anthropology contained the seed of the totalitarianism that we witnessed in former communist states and still witness in the few that exist today.

Popper criticised Marx at length as one of the enemies of an "open society" (a society of which the members are free to criticise politicians and policies, a society that can democratically be changed). If history contains a teleological principle the ideal model of society is determined beforehand. Moreover, individuals are more or less unimportant. In Hegel's philosophy, the individual has egoistic motives; the citizen has a sense of civilian duty, and this thesis and antithesis lead to reconciliation in the state, sensed to harmonise the interest of the community and the individual and to be the expression of true humanity. For Hegel the state was alpha and omega: it *is* the Law, the Moral *and* de Juridical Law. Its only judge is World History. It is divine and must be worshipped: it is the realisation of the human idea, etc., etc. In his *Frühschriften* Marx gave a lengthy analysis of Hegel's conception of the state and rejected it, because he saw through the rhetoric, meant to give the impression of a mystic and deep vision, whereas in the end it came to the identification of the State with the Prussian government. Though to him, like in Hegel's philosophy, the general must prevail upon the individual, his own dream is that when society without classes is realised in the end each individual will be able to enjoy the produced goods, to choose his profession and will have leisure. This is certainly not in se a totalitarian vision. However, till this blissful time, the individual must become conscious of his socio-economic situation and identify his interests with those of the class he belongs to. Individualism is until then a certain token of the wrong (bourgeois) mentality.

It is too simple to see in the position of Marx concerning the role of the individual in the revolutionary process and his stress on the power of the masses, the only source of totalitarianism. The role of the State is crucial. Of course we must specify what we mean by totalitarianism. Let us start by the simple observation that if there is no state there is no totalitarianism. A good way to circumscribe the content of the concept is:

> a political system dominated by a single party and *Ideology*, in which all political, economic and social activities are absorbed and subsumed and all dissidence suppressed by police *terrorism*. Total monopoly of the ordinary flow of information and public argument is essential to such a system....[34]

Plato, Popper's first example of an enemy of free society, did not believe the ultimate goal of history would be reached in the future, but regretted past forms of social organisation. The best form, however, could be discovered by reasoning and was the *final ideal* to be realised. He described it in his famous dialogue misleadingly called *The Republic*. Following the Spartan example, he endorsed communism. Nevertheless, because his ideal state was totalitarian, the suppression of individual freedom he advocated strongly resembles the policies used in communist states to control their citizens. Popper cites the following text from Plato's *Laws*:

> The organization of our forces is a thing calling in its nature for much advice and the framing of many rules, but the principal is this — that no man, and no woman, be ever suffered to live without an officer set over

> them, and no soul of man to learn the trick of doing one single thing of its own sole motion, in play or in earnest, but, in peace as in war, ever to live with the commander in sight, to follow his leading, and take its motions from him to the least detail—to halt or advance, to drill, to bathe, to dine, to keep wakeful hours of nights as sentry or dispatch carrier, all at his bidding, in the stricken field itself neither to pursue or retire without the captain's signal, in a word, to teach one's soul the habit of never so much as thinking to do one single act apart from one's fellows, of making life, to the very utmost, an unbroken consort, society, and community of all with all.[35]

Though Marx would not have recognised his own theories in most of the orthodox ideology of the Soviet Union or in the slogans in the Red Book of Mao, it could be that his work contained elements that could be used to justify State terrorism. Popper believed Marx's thought was profoundly humanitarian and considered his big mistake not his conception of what a human being in essence is, but his contention—which must be seen in the context of the horrible social situations in his lifetime—that politics is powerless in the face of socio-economic situations. This is the real core of totalitarianism: *a democratic state is not capable of bringing change for the good of the majority of people.* His scorn for politics is understandable, but dangerous. In Marx's time, the State was the expression of the suppressive violence against the masses. As a true believer in historicism, he thought the situation could be reversed, if only the State would become, instead of the expression of the dictatorial power of the wealthy against the needy masses of the common people, an instrument in the hands of the latter, in which case it would be able to restore the rights of the majority. This would be only a transitory situation, because once a society without classes would be realised, the State as an instrument for the enforcement of justice would be superfluous. Nevertheless, Marx's dream could be used as an excuse for despotism: the State became the manifestation of the will of the Party, instead of that of the will of the masses.

Notes

1. Fritz Schalk, "Die europäische Aufklärung" in: *Propyläen Weltgeschichte*, vol. 7, Golo Mann, August Nitschke, eds., Propyläyen Verlag, Berlin/Frankfurt am Main, 1986, 469.
2. Cf. for references: M. Gosselin, "Quine's Hypothetical Theory of Language Learning. A Comparison of different conceptual schemes and their logic," in. L. Decock, L. Horsten (eds.), *Quine. Naturalized Epistemology, Perceptual Knowledge and Ontology*, Rodopi, Amsterdam, Atlanta, GA 2000., 2000, 57–77.
3. Cf. J. Hoffmeister, Einleitung des Herausgebers, in: *Phänomenologie des Geistes*, Meiner, Hamburg, 1952, XXII ; XXIII.
4. G.W.F. Hegel, *Wissenschaft der Logik*, Felix Meiner, Hamburg, 1963, 145–146.

5. K. Löwith, *Meaning in History*, Phoenix Books, University of Chicago Press, Chicago, 1949, 55.
6. *Ibidem*, 55–56.
7. K. Popper, *The Open Society and its Enemies*, 3 volumes, London, 1945.
8. Hegel, *Die Vernunft in der Geschichte*, Felix Meiner, Hamburg, 1955, 50–51.
9. Hegel, *Phenomenologie des Geistes*, Felix Meiner, Hamburg, 1952, 244.
10. *Ibidem*, 230.
11. *Ibidem*, 233.
12. Friedrich Heer, *Hegel*, Fischer Verlag, Frankfurt am Main/Hamburg,1955, 39. Heer elaborates upon this aspect, when he compares Aristotle and Hegel: for Aristotle matter time and again disrupts the scheme of the principle of identity [A=A]; it is what contains no logic, cannot be known, the chaotic substratum, the "Mother," that which resists the mind, escapes by fierce opposition the thought of a man, yet is always present and always exists. Hegel does not recognise matter at all, he dissolves it completely in the process of forms, destroys it. "The resistance of the unchanging is completely superseded, broken, annihilated, it is only a transition." Here Hegel falls victim to millenary Platonism and even to manicheanism, which he explicitly fights wherever he encounters it. "Nature is mind alienated from itself."
13. Hegel, *Die Vernunft in der Geschichte*, 55–56.
14. *Ibidem*, 56–57.
15. Hegel, *Phänomenologie des Geistes*, 239.
16. Hegel, *Enzyklopädie der philosophischen Wissenschaften*, Meiner, Hamburg, 201.
17. *Ibidem*.
18. *Ibidem*, 202.
19. *Ibidem*, 204.
20. Karl Marx, *Das Kapital*, Dietz Verlag, Berlin, 3 vols., 1962, VoL. I, 18.
21. Marx, *Deutsche Ideologie*, in: *Die Frühschriften*, A. Kröner, Stuttgart, 1953, 346–347.
22. *Ibidem*, 347.
23. *Ibidem*.
24. *Ibidem*, 358.
25. *Ibidem*, 356–357.
26. Marx, *Das Kapital*, Band I, 185–186.
27. B. W. Clapp, *An Environmental History of Britain since the Industrial Revolution*, London, New York, 1994.
28. Marx, *Das Kapital*, Band I, 186–187.
29. Marx, *Frühschriften*, 361–362.
30. *Ibidem*.
31. *Ibidem*, 353.
32. For the discussion of this interesting question see: E. Bloch, *Über Karl Marx*, Suhrkamp Verlag, Frankfurt am Main, 1980, 145–156.
33. *Ibidem*, 354–355.
34. *The Fontana Dictionary of Modern Thought*, A. Bullock, O. Stallybrass, S. Trombley, (eds.), Fontana Press, London, 1988.
35. *The Collected Dialogues of Plato*, E. Hamilton, H. Cairns, (eds.), Bollingen Series LXXI, Princeton University Press, Princeton, New Jersey, 1961, *Laws*, 942, 1489.

Chapter 8. A complete break with Western tradition

Schopenhauer: the first non-anthropocentric philosophy

Schopenhauer, the successor of Kant

Schopenhauer himself has informed us about the way his philosophy developed: different parts of it were written without pre-established order, they combined, formed an organic totality and eventually a coherent system. The three main sources of inspiration, next to his personal experience, which is in his case primordial for the understanding of his work, were Kant, the Vedanta and Plato, clearly three very different influences, and it is at first sight not evident how he could integrate them in his work.[1] Moreover, the influence of David Hume, though perhaps less prominent, is yet another element of importance. This strange concoction will produce a most original view of the world and of man. Except for the influence on Nietzsche, his ideas have not germinated in the work of other philosophers. The contrast with Hegel, who dominated the intellectual scene, is extreme.

It is well-known that Schopenhauer considered himself to be the true successor of Kant, whose philosophy he had studied and critically examined in order to draw from it the ultimate conclusions that would become the basis for his own philosophy. According to him, the major logical flaw in the theory of Kant was that *das Ding an sich*, the thing as it is in itself when it is not perceived or thought, was presented as a logically necessary consequence of the synthetic a priori proposition that all that exists in the phenomenal world has a cause. Indeed, all these causes needed to be produced by something that had not to be founded in its turn in a further cause, namely by something beyond our experience in time and space. Though, unlike the causes in the phenomenal world, it could not be experienced and therefore could not be known as such, logically this *Ding an sich* as the ultimate cause had to exist. But, the certainty that a reality beyond all phenomenal reality could be demonstrated, is based upon a proposition about the perceptually given reality, namely that everything in it has a cause, and therefore it is also a part of our *representation* of the phenomenal world that is moulded by the particularities of our senses and our mind. In consequence Kant's reasoning cannot with *logical necessity* lead to *das Ding an sich.* Schopenhauer admits that all phenomena we experience in time and space have a cause—he even developed a theory of the sufficient reason in which four causes are distinguished, as an instrument for the comprehension of the world we experience—but he is also certain that neither our intelligence, nor our abstract reason can generate in the best case more than an accurate representation of reality from which cannot be derived an absolute truth. In order to go beyond representation, another faculty is needed, namely *speculation*. It is by direct intuition that we are able to apprehend reality as it is in itself and to discover its true nature: Will. The complement of the world as representation is the world as Will.

Experiencing the world and representing it

Hegel complained that Kant had banned metaphysics from science. He tried in his *Wissenschaft der Logik* (Science of Logic) to remedy this situation and to elaborate a first philosophy founding all ulterior knowledge. Schopenhauer did not start from such a first philosophy, but his views were directly derived from his personal experience of concrete reality. He saw this as an advantage and it made him confident that his ideas were in accordance with the scientific theories of his time, which were also based upon empirical facts. The success of his *Die Welt als Wille und Vorstellung* (The World as Will and Representation) having been meagre, to say the least, he hoped there would be more interest for a substantial essay, where he could explain more clearly the theses of his major work, adding new material to it. Its title is significant: *Über den Willen in der Natur. Eine Erörterung der Bestätigungen, welche die Philosophie des Verfassers, seit ihrem Auftreten, durch die empirischen Wissenschaften erhalten hat. (1836)* (About the Will in Nature. An Examination of the Confirmations by the Empirical Sciences that the Philosophy of the Author has Received). He was convinced that his speculative theory was so firmly rooted in concrete reality that it could be considered the natural complement of science. Indeed, where scientific thought reached its limits it had to be completed by a metaphysics that was its continuation and extrapolation. It is an aberration for a philosopher to neglect scientific data, but it is also an aberration from the part of the scientist to believe science can explain the facts of nature completely. It is just this metaphysical complement his philosophy offers.

The difficulty with this conception is that Schopenhauer could not introduce a single phenomenon without being obliged to derive it ultimately from the eternal principle of reality, the Will. Thus knowledge, commonly contrasted with will, has here to be one of the ways the Will expresses itself in living beings. And indeed, Schopenhauer uses the following beautiful image: "Nature, by means of the knowledge of living beings, lights a lantern to illuminate itself."

Schopenhauer rejected subjectivism and objectivism; there is no subject without an object, there is no object without a subject knowing it. R. Safranski, in his masterly work on the philosopher, comparing the traditional view of man with that of Schopenhauer, writes:

> The picture of man is designed starting from the head. As a rule the reflecting head lets the man about whom he is thinking also begin with thinking. Not Schopenhauer: the "interest" does not start with knowledge, but precedes already knowing and engages us in a totally different dimension of knowledge. "What else is this perceptual world except that it is my representation? (I. 51)," Schopenhauer asks, and he gives the already well-known answer: *Will*.[2]

As explained earlier, the Will is now Kant's *Ding an sich*; it is present in all the phenomena, but in itself undivided, unchanging and eternal, while empirical reality on the contrary is individuated, multiple, changing. Yet, some of its aspects underlying this extreme variety are general, namely the ideal forms after which they are modelled. In fact reality is eternal in its most perfect objectivations, the ideas, the species and the genera, but where the Will is concretely realised in individuals, i.e. under its temporal and spatial aspect, it is imperfect and fleeting. The Will is, as we have seen, not caused and does not cause anything, whereas all phenomenal things are caused, but only by their nearest cause, which means occasionally, not ultimately, and in this sense they remain *contingent*. This means that anything that exists could also not have existed, that it has no necessity whatsoever. As Schopenhauer is an atheist, to exist is not even to be legitimated by being willed by God.

His atheistic view: the world is neither rational, nor divine, nor good

Kant, says Schopenhauer, in his subjective approach, declared that *das Ding an sich* is beyond time, I, in my objective approach, maintain the following:

> … that namely, *das Ding an sich* remains untouched by Time and by that which is only possible by it, coming into being and perishing, and that the phenomena occurring in it could not have that restless, fleeting character, that being near to nothing, if there were not in them a core from eternity. *Eternity* is certainly a concept not based on perception: its content, however, is therefore only negative, as it is the name for a timeless existence. Time is a mere image of eternity, ὁ χρόνοσ εἰκὼν τοῦ αἰῶνοσ, like Plotinus [Enneades, 3, 7, 11] has said: and likewise is our temporal existence the mere image of our essence itself.[3]

This is how Schopenhauer defends the Platonic ideas by reversing the order of things: instead of starting from ideas that have to be posited, he starts from phenomena, and because they are fleeting, in order to exist at all they must have a core that is eternal, i.e. the idea they correspond to.

Romanticism was fashionable when Schopenhauer wrote *Die Welt als Wille und Vorstellung* (The World as Will and Representation), and at that time the internalized feeling for nature, i.e. the emotional approach of nature, was in its heyday. Romanticism was a reaction against the rationalism of the Enlightenment: feelings, sentiments, emotions, had gained in importance and were contrasted with cold reason. The man or woman of feeling became the subject of many a novel; nature was pictured in agreement with the mood of the moment and vice versa: thus solitude found its expression in a desolate landscape, longing in an inviting vista, melancholy in a sombre mountain scene or a dark wood, etc. It was the time of the "Weltschmerz," the suffering that is caused by the world being (alas) as it is and far from ideal. It was also the time of "Sehnsucht," a longing that cannot be fulfilled, of "Sturm und Drang," great passions, heroic actions. In some of the themes in Schopenhauer's work we can recognise a peculiar interpretation of characteristics of the age: it is rather sad that feelings have primacy above reason; reality is, to say the least, quite unsatisfactory and according to Schopenhauer justifies a profound pessimism. He did not share the love for nature of the romantic soul that found consolation in it. He replaced the idealised conception of nature by a more realistic picture of a perpetual struggle for life. The knowledge of human beings, their representation of themselves and of the world is also an instrument of this blind force and brings no relief. *Nothing that we can do can alter the nature of reality.* Because the Will is the primitive ground of all being, and as all our actions are its expression, there is no escaping from it. Also, to the question "why not rather nothing than this world?" there can be no answer, since the world does not justify itself by itself and no reason can be found that would show that the world exists for its own good. In other words, we do not know why Will does not rest eternally in itself, rather than expressing itself in a multiplicity of objectivations. The question is similar to why God, who is perfect, would have wanted to create an imperfect world. The Jews and Christians answer that it is out of his immense goodness. For an atheist like Schopenhauer the answer is not obvious.

His view of the world can roughly be sketched in the following way: Will is the principle of all reality. In itself it is undivided, it is one and beyond the categories of time and space. It is present in its totality in each thing, as much in a rock as in a plant or an animal, as much in one specimen of them as in countless ones. Time and space are the principles of individuation, and in consequence as soon as something exists in these two dimensions, it must be individual. To man it is given to be immediately conscious of the Will, because this is how we, humans, experience our inner life: we are aware of all our emotions, our feelings, the instincts that drive us, as expressions of the Will.

Let us return to his speculation that everything is Will. The Will is not the *cause* of phenomenal reality in its diversity, as there is no action and reaction between both. Though the elements of reality are its individual objectivations in time and space, being of another order, the Will remains unaffected by them. Reality certainly shows regularity, which means that individual things can be grouped together into kinds, corresponding to ideas that form a hierarchical order, *but these kinds do not belong to ontologically different layers of reality.* The reason is the omnipresence of the Will that constitutes the *continuity* between all kinds of things, which is the central intuition in Schopenhauer's philosophy. As we shall see, this profound unity of all the elements of nature will determine his personal, existential way of approaching phenomenal reality. Whereas quasi all metaphysicians are inclined to wishful thinking and desperately want to establish the necessity of themselves and the world by a final ground that by its very nature legitimates everything that exists, Schopenhauer tells us that though we can understand why reality is as it is by recognising in it the expression of the Will, this brings us no consolation, but quite the contrary, it brings us desperation. If we consider his thesis in the context of the succession of world views in the history of Western philosophy, we can measure its originality. Though Schopenhauer too posits a first principle beyond space and time, a principle that is one and indivisible and of which all things are the expression, his position is unique: the Will is *neither divine, nor rational, nor good.*

Fundamentally the elements of reality are all one, but as we know from the foregoing, their forms are different. The variety of individual realisations of the Will is modelled upon ideas, and though the individual elements are ephemeral they are succeeded by ever new ones that are instances of the same eternal forms. This means that nature consists of a series of models represented by populations of particular concrete realisations, ranging from the inorganic to the organic, in the realm of the living beings from the plants to the animals and in the range of the animals from the lowest species to man. (The Will cannot be identified with Life. Though inorganic matter is the expression of the same principle as living matter, Schopenhauer denies that this means that life already pre-exists in the former.) Because of this continuity, one can learn as much about the world by looking at one individual of a kind than if one could travel to the limits of the universe and study all those one encounters during the voyage.

The will to live and to procreate is present in all living beings; the will to survive is so strong that this goal is pursued at all costs and especially at the cost of other living beings. Death and life are complementary aspects of the same principle. All animals are capable of suffering, but man, knowing about himself, is the animal that suffers most. The Will as the foundation of reality is without purpose, without knowledge, a blind force, a perpetual compulsion, the source of endless sufferings. Christians believe that we must contribute to the coming of the Kingdom of God on Earth. Saint Augustine explains that the sense and meaning of profane history is to be sought in sacred history that will end with the achievement of this goal, Hegel considers man to be the instrument for the realisation of the "*Weltgeist*" and Marx is convinced that we are the instruments for the creation of a world without classes. There is in the Western tradition a profound belief in an ultimate aim in history. We, humans, whether we know it or not, have a mission, we are necessary for the completion of reality that is still unfinished. Schopenhauer has no such historical and progressive view of reality. He declares on the contrary that our sufferings do not make sense:

> If the world was not something that, practically speaking, should not exist, then it would also not be a theoretical problem, but rather would its existence either require no explanation at all, as it would be so completely understandable in and by itself that an astonishment about it and a question about it could not arise in any head; or its purpose would unmistakably appear to us.... Therefore, if somebody would dare to raise the question why there was not rather nothing instead of this world, then it must be answered that the world cannot be justified by itself; no reason, no final cause of its existence can be found in itself, and it cannot be proven that it is there for itself, i.e., for its own benefit.[4]

If the reality has no hidden meaning, human existence has no meaning, and as we cannot change the world, we have no mission.

We are our bodies and know them immediately

Safranski puts it thus:

> The Will is what is most certain. "Will" is the name for the self-experience of one's own body. The only reality that I do not have as a representation is my body that I *am* myself. However, because I am at the same time able to relate myself to my body by representing it, my own body is given to me in two ways: once as representation in intelligent perception, as an object amongst the objects...; but also at the same time in quite another way, namely as that what is immediately known to each of us, as that what the word Will means.[5]

On some privileged moments, when in a speculative mood, a man can free himself of the pressure of the Will and can contemplate its working in nature:

> He who has emptied himself from the evil urge of the Will is free to see in everything only the spectacle of it. The main actor of this spectacle is the body. Schopenhauer's philosophy of the body sweeps away the traditional body-mind dualism, and takes on something that was unheard of: the body as the corporeality of the will becomes the founding principle of a whole metaphysics.[6]

Founding principle, yes, but not something to worship; this grumpy philosopher never liked the body and its instincts, let alone that he would have wanted to glorify it. For such a positive view we shall have to wait till Nietzsche.

In order to understand the novelty of Schopenhauer's conception we must refer to the idealist tradition. Schopenhauer rejects the difference between the acts of the body and the acts of the pure will that was the presupposition upon which was founded the mind-body dualism of Plato. We must note that for the latter two kinds of will existed. The first kind of will was a purely spiritual impulse residing in the mind and reigning over the body. The second kind, by contrast, was willpower originating in the body, which is a power of non-being, of death. The body is a grave, but the pure, immaterial will can overcome the tyranny of the bodily impulses. Western tradition is full of attempts to make the body and its instincts vanish or at least to make it a negative kind of reality, and this always presupposes that the headquarters of a man are in the soul, the mind dispatching orders to the despicable body.[7]

It is because the Will expresses itself in its individuated objectivations, that living beings *are* their bodies instead of *having* them. It was never doubted that animals coincide with their bodies, but the novelty is that this holds also for men, even though they are conscious and self-conscious, and this coincidence constitutes the continuity between all living organisms. This philosophy contains a most curious idea: *continuity between man and animals, without there having been evolution.*

We have no free will and no mission

To be capable, like the Christian mystics, to transcend, in a moment of grace, individual existence and to be reborn beyond individuality, is the only possibility to be free. However, this state of blessing is only given to exceptional individuals. Of course animals lacking abstract reasoning cannot resist their natural needs and have to follow their basic instincts more than we do:

> The animal does not have the possibility to be free, as it is without the possibility of a genuine, i.e. considered, choice after a preceding total conflict of motives, which would have to be for this abstract representations. Therefore, with the same necessity with which a stone falls to the earth, the wolf puts its teeth in the flesh of its prey, without the possibility to understand that it is itself as much the eaten as the eater. *Necessity* is *the Dominion of Nature*, *Freedom* is *the Dominion of Grace.*[8]

We are not free by the power of our intellect, but by being able to reach a special condition in which we can see the true nature of reality that is one.

Schopenhauer declares that the existence of an immaterial soul is a mistake made by Plato, who cannot accept that the will and the intellect have their cause in the body and who therefore posits as our true self an immaterial entity, an eternal soul. In other words: the soul is nothing but the hypostasis of the cause of our mental processes.[9] To Schopenhauer it is clear that like it would be absurd to seek the cause of the digestion of food elsewhere than in the stomach, likewise it would be absurd to believe our will and thoughts are different from the functioning of our brain. No free will can exist if our behaviour is a product of an organ of our body and is determined by it. He does not deny that we can decide after having considered various possible choices and their implications, but this rational decision is not an act of the will in the true sense, because an act of the real will always originates in the body. Only, we are more often than not fooled by ourselves and believe we act rationally, while we are following the impulses of the will without knowing it. The Earth is a vale of tears and nobody escapes the sorrow and suffering caused by the working of the Will. Though we have the impression to be free, we are not: we coincide with our body, we cannot escape its urges, especially not the sexual urge that is the strongest of all, because it serves reproduction and here too we are not different from the animals. The perfectibility of man, his freedom to change himself, to choose his destiny, to alter history, all this is illusion. We cannot change the character, the personality we are born with. Schopenhauer, who admired his father as a morally good, upright man, with noble principles, had often painfully been conscious of his own shortcomings, and of his fierce temperament. He had tried hard to change, to discipline his urges, but in many cases lost the battle and this must have contributed to his pessimism concerning himself and his congeners.

For thousands of years a human being was human because he was a person, which meant that he was a soul temporarily inhabiting a body. Hegel's conception is that the most outstanding characteristic of human beings is that they are not merely bodies that feel and think, bodies that are there, that are *an sich*, which is the non-reflective mode of existence of the beasts, but that they are *für sich*, i.e., that they are there for themselves. Men are conscious and self-conscious; they know about themselves and can distance themselves spiritually from themselves, act upon themselves and upon their situation. They are essentially freedom, and in consequence men are historical beings capable of modifying reality. Marx has a kindred conception; though the reality of a man is determined by his economical and social context, individuals are able to become conscious of the nature of the economy and the society they are part of, to grasp their mechanisms and their significance. They are able to unite themselves with others in order to change the existing political and social order and in the end to shape the nature of humanity. If the individual must make sacrifices, even give his life for the cause he supports, this must be seen as a contribution to the realisation of the ultimate aim of history. Schopenhauer's picture of man is completely different. Though on privileged moments some of us can forget about the tyranny of the impulses of their bodies, in daily practice we cannot escape their impact. Even if a man becomes aware of the true nature of reality, even if he becomes a saint leading an ascetic life, even if he is aware of the continuity in reality and feels compassion for the whole world, even if he is an aesthete recognising the eternal forms in the fleeting phenomena, he can only momentarily untie the shackles of the Will. Human freedom and historical mission is a deception. As long as a man lives he remains an expression of the eternal Will, even if he thinks otherwise, and when he dies others will take his place: "*Natura non contristatur.*" We can learn from Indian mythology, especially from Brahmanism, that life and death are one. History is merely a by-product of the eternal cycle of coming into being and falling apart. The Westerner is unaware of this; for him there is an arrow of time of which the course cannot be reverted. After a historical event nothing will ever be the same. The choice by an individual of a line of action is essential and his acts can make all the difference.

Schopenhauer's behaviour was in line with his convictions. His indifference to historical events and the refusal to take any part in them is clearly illustrated by his attitude during troubled times for his country. Prussia was part of the Napoleonic empire and contrary tendencies were to be witnessed in the population; for most Germans patriotism and pragmatism were to be kept in balance. In academic circles, in spite of the censorship by the authorities, politics became the hot topic: should action be taken and Prussia regain its independence? Many of his colleagues wanted to do something concrete and took up arms. Meanwhile the "Grande Armée" assembled in Berlin and the campaign for the conquest of Russia began. It was a disaster for the French and when the remnants of the once so valiant troops entered Berlin the spectacle of the misery of the soldiers could hardly be imagined. Schopenhauer is a direct witness of the tragedy that offers the opportunity for action, but he declares that patriotism is not his cup of tea and that his country is broader than Germany. He reflects on the misery of individual existence and on death as it is the fate of each individual, but he is not interested in History. His subject is reality in general and not successions of peculiar historical processes taking place at a specific time and place and that are the result of chance elements. His only preoccupation at that moment is to write the work he is "pregnant with" and feels grow in him, like a good mother feels the growth of a child in her womb. He flees Berlin and seeks a quiet place to give birth to his *opus maior*, his major work.

Schopenhauer's theory of the species

Empirical knowledge implies that each of its propositions about facts is validated by four causes to be compared with the four causes of Aristotle: the reason of change, (the efficient cause), the reason of being (the material cause), the reason of knowledge (the formal cause), the reason of acting (final cause). The reintroduction of a formal and a final cause opens the door for a teleological approach of biological phenomena. Once science has studied the reasons for all the phenomena it investigates, once it has generalised its results and formulated laws of nature, its task is finished, and, as we have seen, philosophy must supply speculatively the revelation of the true nature of the reality behind the phenomena, their so-called "noumenal" aspect.[10] This is especially necessary in the domain of Life, but the task of philosophy can only be complementary to the scientific enterprise, if a correct method is applied. Schopenhauer rejects the analytical approach of Descartes and of other materialists, because it presupposes that we can explain each thing by reducing it to its material parts, and these parts to their parts, till the atomic level is reached, and then by specifying the mechanical causes of movement and change. Indeed, the analytical method is insufficient, certainly where living beings are concerned, because, if it was correct, this would mean that the different forms of life are the result of the interplay of natural forces, which cannot mean anything else than that they are created by chance. In fact the physicists have chosen matter as *Ding an sich*. However, as Kant has stressed, though Newton has shown how, given the existence of matter, the physical world can be explained, we cannot hope to find a "Newton of the sprig of grass" explaining the coming into being of living beings in the same way.[11] While Schopenhauer rejects a teleological vision of history, he advocates teleology in biological matters.

Though even in inorganic matter the blind and dark driving force of the Will is present, this force is more clearly expressed in the first primitive living beings and in the plants. It is present in the lowest animals, like the molluscs, though no conscious goal-directed action can be found in them, but only movements caused by stimuli and the functions necessary for survival controlled by a simple nervous system. In more complex animals that have to transport themselves in order to find food and mates, the Will is expressed in their way of life that makes it necessary for them to have sense organs for gathering knowledge about their environment. In man, at last, the Will realises its highest form. From the moment a well developed nervous system is present, the Will is "represented" by a special organ, the brain. The brain is a mechanism, a tool that after the world existed only as *Will*, produced the world as *Representation*.

Schopenhauer particularly admires those biologists who are morphologists, collecting specimen, comparing them, ordering them, systematising them. They give us clear knowledge that enables the philosopher to speculate about the different objectivations of the Will moulded upon the ideal forms. We can see in the vast assortment of these forms that life is not a mechanical process, but a *force*. The reverse would mean that the Will does not manifest itself in qualitatively different objectivations, but everywhere in the same manner, with only quantitative differences.[12] Thus Schopenhauer's Platonism makes him return where life is concerned to Aristotle, who was an excellent biologist, but never totally freed himself from the influence of his master Plato. Though the world consists only of substances, particular concrete things that can exist autonomously, the natures, the species and genera, when not stripped unequivocally from their ontological implications, are perfect and eternal models. Combined with his teleological approach, the essentialism in Aristotle's work has hindered the progress of science, until finally his metaphysics was discarded in most domains. In biology the emancipation from Aristotelism came late, with La Mettrie, then with Lamarck, and finally with Darwin. Schopenhauer, however, resolutely chose the side of the Aristotelians.

Life itself as the expression of the Will is nothing but the general warfare of living beings against other living beings in their struggle for their material subsistence. This view came very close to evolutionism, but his Platonism prevented him from drawing the right conclusions. He took much trouble to master and assess the biological theories of his time though, but he was bound to choose the wrong side, or rather, the side of that history would show to be in the wrong. Here we witness his betrayal of his own principle that his metaphysics must not be a priori, that it had not to be his starting point when assessing scientific data. In fact to him the existence of eternal species and genera was so certain, and this could be so smoothly used to interpret the results of the most famous biologists of his time, that he had no doubt he was right.

Though the Will is equally expressed in all the forms of life, as we have mentioned, Schopenhauer represents the different species as hierarchically ordered. He gives a metaphysical explanation for this order:

> When of the realisations of the Will on the lowest level of its objectivation, i.e. the inorganic realisations, a number enter into conflict with one another, because each of them wants to get hold of the available matter following the laws of causality, then from this struggle results a higher idea, which overpowers all the other less perfect ones that were already present before it, but in such a way that it lets their essences exist in a subordinate way, by absorbing an analogue of them. This is a process that can only be understood from the identity of the Will in all the ideas and from its striving towards ever higher objectivation....[13]

The species constitute a hierarchy, but not a progressive evolution. It must be clear that in this, the coming into being of the different forms of life, there is no temporal succession. Indeed, he reproached Lamarck that his theory implied that all the forms of life evolved over time and in consequence in spite of and their extreme diversity had to stem from one primitive living being, which seemed completely absurd to him. His vision, he claimed, did not contain this flaw, but was in accordance with that of the famous Cuvier, who, though or rather because he was an excellent scientist, specialised in comparative anatomy, remained an Aristotelian, true to the classical idea that the species formed an eternal rational order. It is also evident for Schopenhauer that the coming into being of new forms of life cannot be explained physically and chemically alone, but must be understood as the production by forms of ever new forms.

Whoever adheres to the theory that eternal models, natures, essences exist, transcendently or immanently, cannot accept that something radically new can come into being, and therefore cannot accept the evolution of species. This is what induced Schopenhauer into his error of judgement. He explicitly rejected the idea that a rational model, a divine plan, preceded the order that we discover in Nature. As he repeatedly stressed, Nature produced the intellect and not the other way round. Will can be witnessed everywhere, it is primitive. Therefore, he declares, all teleological facts can be explained by the Will in the beings in which its working is observed. Organisms are constructed in such ways that all their parts, all their organs are just right for them, because they correspond to the particular way in which the Will expresses itself in their form of life. Their behaviour does not follow from the way they are constructed, but rather the other way round, the way they are constructed follows from their behaviour. The opposite would leave the question open why all the parts and organs are geared with one another, and why some of them would not be more appropriate for another life style than the one we observe, while only the most important would have to be in accordance with the one they actually have.

That the way of life must be the reason for the structure of an animal and for *all* its parts, can be derived from the fact that a carnivore has strong claws to catch its prey, and has also the right teeth to tear it apart and shred its flesh, that it has the matching short digestive tract and the right limbs to enable it to persecute its quarry and pound upon it, that none of its organs seems to be superfluous.[14] He gives numerous examples to prove his teleological approach is the only possible one; like Aristotle contended, the formal cause is primary, it is at once the essence and the aim of the animal and determines its construction. Those zoologists who do not understand the metaphysical ground for this cannot fail to make a regrettable mistake:

> This truth imposes itself on the reflecting zoologist and anatomist with such an evidence that, if his mind is not purified at the same time by a deeper philosophy, he can be induced to curious errors. This is what has indeed happened to a first class zoologist, to the unforgettable *de Lamarck*, to him who has won for himself an immortal merit by the so profoundly grasped division of animals in vertebrates and non-vertebrates. In his *Philosophie Zoologique*, Vol. I, cap. 7 and in his *Histoire naturelle des animaux sans vertèbres*, Vol. I, Introduction, pp. 180–212, he declares in all seriousness and tries to explain extensively that the form, the proper weapons and the organs for action in the external world of all kinds, has not at all been already present in each species from the beginning, but *originated* only as a consequence of the wilful purposes of the animal, determined by the features of the situation and of the environment and by its own repeated efforts and the habits created by this, and this slowly *in the course of time* and by continued generation.[15]

Then follows an enumeration of examples given by Lamarck and he concludes ironically that Lamarck forgets that the process must of necessity take a long time, and that if the evolving species were not well-equipped from the start they would have become extinct, (a remark that was also made later on about Darwin's theory of evolution). Like Kant, Schopenhauer denies that time and space are real and not ideal and he continues:

> So blind makes a hypothesis once conceived. But this one here arose from a very right and profound conception of nature; it is an error of genius, which, in spite of all the absurdity that lies in it, honours him. The truth in it belongs to him as a naturalist: he saw rightly that the will of the animal is the primary and has determined its organisation. The error, on the contrary, is due to the backward situation of metaphysics in France, where Locke and his feeble follower Condillac still reign and in consequence bodies are things in themselves [*Dinge an sich*] and where time and space are properties of these things in themselves and where the doctrine, great and rich in consequences, of the ideal character of space and time, and thereby of all that arises in them, has not yet succeeded. Therefore could de Lamarck not but imagine his construction of creatures as in time, by succession.[16]

Kant was the first philosopher to understand time and space as categories of the mind, not dimensions of matter. Schopenhauer concludes that if time and space are no longer ontological elements, the species cannot come into being in a spatio-temporal process, but must be eternal forms. In Germany such a mistake would not have been possible after Kant had made clear that all knowledge is a representation of reality and that neither the simplistic and absurd atomism of the French, nor the reduction of reality to the physico-theological of the English were acceptable.

Nature: perpetual suffering

Time and again Nature offers us cruel spectacles; the individual does not count, only the conservation of the species. Schopenhauer uses as example of this the ichneumon wasp that deposits its eggs in caterpillars; it is even known that its larvae when they hatch eat the flesh first and leave the nervous system intact for the caterpillar to remain alive as long as possible and in order that it does not decompose too rapidly. He also mentions polyps growing on the mother polyp and still fastened to her with their stem fight for the food of the mother. The examples cannot be terrible enough: he cites the Australian Bulldog ant he has read about in a journal, of which, when it is cut into two, the halves fight with each other, the anterior part using for the battle its mouth parts, the posterior half its sting, till death follows. Plants, though the most innocent of the living beings, because generally they do not feed upon other plants, occasionally are parasites. Or they grow upon other plants like the wild vines strangling the oaks they use as support.

The Bible tells us that God created the world in seven days, and that at the evening of each day, he saw that his work was good. In Schopenhauer's philosophy Nature is not the work of God; for the first time Nature is depicted as a terrible force, not to the measure of man and not held in check by a divine being. Though what goes on in it does not correspond to and even is contrary to his moral standards, man is part of this reality. The terrible fact is that Nature is a mirror in which he can see his reflection and he cites again Plautus:

> Thus we see everywhere in Nature struggle, battle and alternation of triumph [and defeat] and we shall in this again recognise more clearly the essential splitting of the self from the self. Each step of the objectivation of the Will tries to deprive the other of its matter, space, time. Continuously the remaining matter must change its form, as following the rule of causality mechanical, physical, chemical, organic phenomena eagerly try to supplant each other to occupy the foreground; they dispute each other their matter, because each of them wants to reveal its idea.... Most clearly visible is this general struggle in the animal world that feeds on the world of the plants and in which each animal itself becomes the prey and food for another one. This means that the matter in which its idea realises itself, must abdicate for the realisation of another animal, as each of them can only safeguard its existence by the elimination of another one, in such a way that the will to live constantly lives of itself and under different forms is its own food, till mankind, because it usurps all the rest, considers nature a product for its own use, and this same species also ... discloses in itself this splitting of itself and discloses "homo homini lupus" [Man is a wolf for men] with the most terrible obviousness, (Plautus, "*Asinaria*" 2, 495).[17]

The idea that man is egoistic, and this in spite of his lofty principles of altruism, is in se not original, but in Schopenhauer's work it is not a general observation, but a direct consequence of his fundamental world view and has therefore a more profound meaning. New is that he denounces anthropocentrism that sees nature as subservient to man and gives him a kind of licence for its limitless exploitation, while hitherto this had not been a subject for reflection.

Man and animals

All animals have intelligence; even the lowest species are at least able to understand the causal relations between themselves and the outer world, while the higher animals are also able to grasp the causal relations between the things and events they observe in their surroundings. Man is not only intelligent, he is rational, i.e. he is capable of abstract thought and even contemplation. The Will here uses an extra tool, but this tool creates for human beings the possibility to oppose the Will, to have the intuition of ideal forms instead of seeing only the concrete individuals. It becomes possible to break loose, if only momentarily, from the obscure forces of the will he is subjected to.[18] It is remarkable that Schopenhauer believed that in Nature Platonic ideas were realised and at the same time was convinced that, because they are objectivations of the same Will, there is a continuity between men and the rest of nature. He has been said to like animals more than men and it is not surprising that he has been represented several times walking with his little dog. Like Hume, who also liked animals, he believed their intelligence was an evidence. Intelligence is the same in animals and men, he says: it is the grasping of causal relations. By this standard some animals are extremely intelligent, whereas it must be said that some men are extremely stupid. Most stupid are those who are not capable of empathy. They are not capable to step into the shoes of another person, to imagine what they feel, want, think, and many of them do not understand the hidden motives behind the behaviour of others. As an extreme example of the inability to recognise causal relations, he mentions an idiotic boy in a psychiatric clinic, about eleven years old, who was not completely devoid of rationality, as he could speak and understood what was said, but was less intelligent than many animals. Each time Schopenhauer came, he was fascinated by the eyeglass the philosopher wore on a string around his neck and that reflected the window and the treetops outside. He marvelled at this reflection, but did obviously not understand from where it came.[19]

Schopenhauer was in advance of his time by his explicit rejection of anthropocentrism, but also by his progressive ideas about the capacity of animals to suffer. If there is continuity between all living beings and in consequence between animals and men as the expression of the same transcendent principle, those who have understood this cannot but condemn cruelty against animals. Though he was an adept of quietism, on behalf of the animals he used his sharp pen to denounce the horrors of vivisection. Believing white men needed meat to be healthy, he was not a vegetarian, but demanded that animals, if they had to be slaughtered, had a right to be put to death as painlessly as possible, under sedation with chloroform. Descartes carried the distinction between matter and soul to an extreme and declared animals to be machines without soul. (He probably never had a dog or a cat). He was not only wrong, which is forgivable, as we all are liable to this condition and we have to put things in a historical perspective, but denying animals a soul and not only the capacity to think, but also to suffer was crass. Being for a long time a very influential philosopher and scientist, he caused much misery for our fellow-animals and this for centuries. Schopenhauer should be honoured for his plea for compassion for other living beings.

Remedies for suffering

We can not alter the nature of reality, we cannot escape our own nature, but we can forsake temporarily our individual viewpoint and egoistic motives. Another relief of the pressure of the will that drives us and drags us along can be found by practising ascesis that is one of the remedies Schopenhauer proposes. Indeed, once we have satisfied our urges, reached our goals, the temporary satisfaction is soon over, we are bored, we seek new pleasures, set ourselves new aims, and suffering starts all over again if we do not forsake our primitive desires. In the bleakest despair about the human condition we can consider suicide, but this does not alter our sad condition; on the contrary it is also an expression of the ever-present Will, eliminating an individual, but continuing its working in its congeners. A final consolation can be found in art: exceptional individuals are able to create works that render the ideal aspect of reality, grasp the most perfect objectivations and their underlying principle. He prefers music to all other forms of art, because it is not a representation of ideas, the eternal essence of the phenomenal aspects of the Will, but a direct expression of it.

Thus we share with the animals not only our intelligence, but our instinctive and emotional life. In one respect, however, we are different: we have not only the capacity of abstract thought, but also of speculation. This means we are able to grasp the principle of reality and how it is realised in its objectivations. We can become aware of the profound unity of all the elements of nature, witnessing the abject cruelty of wildlife, the blind forces at work in plants, the blind instincts dictating the behaviour of animals. The strongest desire we have as individuals is to ascertain our own survival and that of our offspring. It is not, as the Romanticists contended, a supreme pleasure to merge into nature, to let go of our individuality, because this goes against our animal nature. Yet, the privileged moments exist, when a man is able to be free of will and to be morally good:

> He becomes conscious of the fact that the difference between himself and others, which to the evil is such a deep abyss, belongs only to a passing, deceptive appearance: he recognises, immediately and without reasoning, that in itself his own appearance is the same as that of the other, namely that will to live that is the essence of each thing and is present in everything. Even more, he is conscious that this will covers the animals and the whole of Nature: therefore he will not torture animals. He is now as incapable to let somebody remain in need, while he himself has what is superfluous and unnecessary, as [to tolerate] that somebody would be hungry for one day, to have more than he can enjoy the next. For to whom accomplishes acts of love the veil of Maja has become transparent and the deception of the *principium individuationis* has left him....[20]

This universal compassion is a consequence that has to be drawn from his metaphysics: if in nature all that exists is the expression of the same first principle, then we recognise ourselves in all its elements and are in sympathy with them. Nietzsche saw in this universal compassion a remnant of Christian culture and its precept to love one's fellow-men, but Schopenhauer clearly believed it was much more part of the Hindu world view that, as we know, inspired him. As mentioned, he often denounced and condemned cruel behaviour against animals, and expressed his indignation. He clearly believed that its source is Christian religion:

> These are the effects of the first chapter of the book Genesis and of the whole Jewish conception of nature. For Hindu and Buddhists the "Mahavakya" applies (the great word: "Tat tvam asi"—this are you) that at each moment must be said about each animal, in order that the intrinsic identity between it and us be present to our mind as a guide for our behaviour—so far for your most perfect morality![21]

Conclusion

Schopenhauer's metaphysics is unique, because it posits as first principle of reality the Will that is not justified by itself, but *aimless* and *senseless*. The West had never before recognised the lack of meaning as the fundamental characteristic of reality. Like religion, metaphysics is supposed to correspond to the unquestioned universal need of human beings to make sense of their own existence, of their place in the world and of the world itself. The ineradicable wishful thinking of men forbids them even to consider the possibility that beyond the world they experience there is no transcendent reality founding it. Schopenhauer does not deny that there is more to reality than its phenomenal aspect and more than science can tell us, but what he discovers does not correspond to what deep in their hearts most men wish. The world we live in is not the best possible one, but rather the worst possible. He is famous for his dictum "the worst has still to come." He denies that the world has a justification for its existence in itself, and as an atheist he can neither believe that it has a justification in a Demiurge who made the primitive Chaos into a Cosmos, nor in God the Father who created it and not even in the more abstract God of the Deists.

Safranski, after other biographers, shows that Schopenhauer, as a child, felt he was not loved by his mother and he refers to this psychological element as the main explanation for the philosopher's gloomy and pessimistic view of reality. Indeed, it must be said that Charles Darwin was equally aware of the struggle in nature, of suffering and death as a normal component of life, but he weighed this against the beauty and happiness that are equally present in it and he came to the conclusion that the balance did not necessarily tilt to the negative side. Though Schopenhauer does not deny that there are good moments, to him the moments of sorrow and suffering outweigh them by far.

Nietzsche's philosophy: breaking with a millenary tradition

In order to complete the panorama of the philosophical conceptions of man in the nineteenth century, I have chosen Friedrich Nietzsche. In the previous section, I have sketched the outlines of Schopenhauer's anthropology and tried to show its revolutionary aspects, against the background of the common hierarchically layered ontology of the Western tradition. The opposition could not be sharper than between Schopenhauer's anthropology and those of Hegel and Marx, who both had a teleological and progressive view of reality. They saw man as the elected instrument for the fulfilment of the ultimate promise history contained. Schopenhauer's philosophy was quite unique, by making of man a part of Nature and the playing ball of its driving forces. Peculiar in his philosophy was the combination of rational radicalism and pessimism with Platonism. Though to him it may have seemed that he had integrated these elements in a coherent system, the combination remained unsatisfactory, and it prevented him from drawing some ultimate consequences that imposed themselves. Nietzsche was in his younger years an enthusiastic adept of Schopenhauer, and though later on he rejected the idealism and pessimism contained in his philosophy, he shared an important number of radical views with him. He continued the work of Schopenhauer breaking down the last walls of the millenary philosophical tradition that still stood erect.

In Nietzsche's life we must distinguish different periods and this is of great importance for a correct understanding of his ideas. If we do not take this into account and quote him out of context, we can interpret his ideas anyway we like. This was often done by philosophers who liked to represent him as a forerunner of them. Unlike many other philosophers he did not write a major work and rework later on the salient themes, but his opinions several times completely changed in the course of his life. These changes were caused by the encounter of men of significance, such as Richard Wagner and the philosopher Paul Rée. Nietzsche was also a great reader, and consulted books in the most diverse domains of which a number had a lasting impact upon his thought. As becomes clear from the *Nietzsche Kronik* (Nietzsche Chronicle)[22] made by Karl Schlechta, his interests did not only pertain to the humanities, but to the natural sciences as well.

The influence of Schopenhauer

Nietzsche was from early on interested in philosophy. He was brought up in a period when romanticism and German idealism were still in their heyday. As an adolescent he was trained in the humanities at the Gymnasium of Schulpforta. As the name of the school indicates, the ideal was an all-round humanistic formation. He compensated for the discipline and rigor of his studies by reading romantic literature in his spare time, by listening to romantic music and discussing with his friends the cultural events of the moment. He was conscious of the too great disparity of his interests and of his emotional instability, and he felt uncertain about the right choice to make for his future. Eventually he chose to study, next to theology, which was inevitable for the son of a minister, classical philology, which he supposed would secure him a certain intellectual order and discipline. He started his studies in Bonn, but soon followed his professor, Friedrich Wilhelm Ritschl, to Leipzig. Against the wish of his mother and aunts, who had taken the responsibility for his education after the death of his father, he discontinued his studies in theology. While browsing in an antiquarian bookshop, he discovered Schopenhauer's *Die Welt als Wille und Vorstellung* (The World as Will and Representation).[23] This chance event would determine his future life. He was fascinated by this work, which in many respects was so extraordinarily different from the contemporary views about life, human existence and moral values. From his letters we know what this philosopher has not only intellectually, but also emotionally, meant to him in this period of his existence.[24] In short, Schopenhauer became his *maître à penser*.

Being a brilliant student, he was recommended by Ritschl for an academic career at the University of Basel. He became professor when he was only twenty-four years old. A fervent music addict, playing the piano and composing, he liked to associate with cultivated people who shared his passion. It was through the wife of a colleague that he was introduced to Wagner, whom Nietzsche already when he was an adolescent admired as the leader of the avant-garde in nineteenth-century music. The deep friendship that grew between the young scholar and the famous composer, when they discovered their mutual admiration for Schopenhauer, consolidated Nietzsche's infatuation with the latter's romantic and pessimistic philosophy. As we shall see, Schopenhauer's vision of reality has been no doubt the main source for his own philosophy. As is often the case, a young adherent will at first eagerly absorb the ideas of his predecessor, then, becoming more critical, strip them of what seems obsolete to him, but retaining the core of what is true and useful in them, and finally drawing new, stronger conclusions. This core consisted of the atheism of Schopenhauer, his contention that man *is* his body and thereby is a part of nature, that man has no free will, but is ruled by his instincts, and finally that the world we live in is not divine, rational and good and that man is not its centre. Throughout the different stages of his life, in spite of the fact that they testified of deep changes in his approach of reality, abandoning romanticism for an idiosyncratic form of positivism and then creating a unique type of philosophy, Nietzsche will remain faithful to these fundamental ideas and they will mould the view of man that he will develop.

Paganism and positivism

Presocratic philosophy remained a complementary source from which Nietzsche would draw his ideas. He always tried to find a good balance between his scientific work and his general interests. When he was still very young he needed the rigor of philology to discipline his romantic mind. When teaching in Basel, he sought on the contrary to compensate for the all too strict discipline needed in his scientific work and his lectures. He refused to become a narrow-minded *Bildungsphilister*, a philistine only interested in philological texts. His first book, *Die Geburt der Tragödie aus dem Geiste der Musik* (The Birth of Tragedy out of the Spirit of Music), written at the time of his friendship with Wagner, was proof of the broad scope of his thought. Inspired by the metaphysics of Schopenhauer, he developed in this work the idea, which was extraordinary at that time, that in Greek culture the Dionysian drive of our instincts and the Apollonian rationality are complementary and of equal importance. Dionysus was the God who was torn to pieces, symbol of the primitive One, scattered in individual existences of the phenomenal world, where Apollo reigned. The cult of Dionysus was associated with the wild rites of a northern barbarian religion, but Nietzsche had the more refined Greek veneration for the God in mind. However, this conception of the culture of the ancient Greeks comprising two opposite fundamental elements was so revolutionary that it discredited the young professor in the eyes of his more conservative colleague philologists.

During his studies he had worked on pre-Socratic philosophers of nature, Thales, Anaximander, Heraclitus, Parmenides, Anaxagoras, Empedocles and on Democritus. Perhaps the study of these authors made him immune to the influence of idealism. In particular the Heraclitian conception of reality impressed him: nature is the ever-changing result of the interplay of forces, engaged in a continuous struggle, tugging at each other in different or even opposite directions.

To the ancient Greeks, the world is eternal, not created. It is Chaos ordering itself into a Cosmos. This view of an uncreated world must have seemed attractive to Nietzsche, who as a young man had slowly lost his faith, without going through a crisis. Curt Paul Janz, his outstanding biographer, suggests that Nietzsche did not have a religious temperament. Following Janz, even as a child he could not accept the central Christian concept of original sin. He adored his gentle, cultivated father and could not believe that such a good man could be tainted by it.[25]

The archaic Greek society described by Homer corresponded to an aristocratic ideal Nietzsche had forged for himself. He was born in Röcken, near Lützen, in rural Prussian Saxony. His father, Ludwig Nietzsche, and also both his grandfathers were Protestant ministers. One of them, grandfather Oehler, was more a country gentleman than a clergy man. Moreover, his father taught at the local court and was a fervent royalist. Nietzsche was brought up in this milieu dominated by the country nobility and this left a lasting impression. He even believed the family legend that the Nietzsches were descendants of Polish aristocracy. No wonder that he recognised his ideal in the aristocratic heroes of the archaic rural society described in the Greek literature he had studied. He remained his whole life a conservative and being an individualist, he dreaded above all the rule of the masses. His individualism went together with the ancient anti-idealist thesis that only what is particular exists apart from the mind.

Nietzsche had an independent and critical mind. He had been for a while member of the "Schopenhauer Society" and had propagated Schopenhauer's philosophy. He had also defended Wagner and his artistic conceptions in his non-professional writings, but it became soon clear to Wagner that intellectually Nietzsche would go his own way. It could be foreseen that he would not follow unconditionally the Maestro in his opinions. Nietzsche, moreover, looked with suspicion at the admirers and flatterers surrounding Wagner. He was especially wary of their cult of the German past, of German mythology, of the stupid nationalism that was paraded, and he strongly opposed the virulent anti-Semitism of some of them. Once the most fervent defender of Wagner's creations, now he visited Wagner's home at Tribschen less often and showed a certain coolness that did not escape Wagner's attention. This could not but end in a final separation. Nietzsche did not attend to the creation of *Parsifal* in Bayreuth, because the Christian message the opera contained was too obvious, and because, in his opinion, the music was intended as a kind of opiate for the public. Wagner had borrowed many ideas from Schopenhauer for his theories about music and about what he called *Gesamtschauspiel* (opera as a performance combining plastic art, and theatrical and musical elements). Nietzsche reproached Wagner that in his opera's the pagan spirit was lost and that he was preaching Christian virtues, ascesis and compassion. Nevertheless, he genuinely loved Wagner and his wife Cosima. Also, the inevitable breach that had grown between them was a personal drama for him, but he simply could not play the hypocritical role of the obedient follower. This was also the end of his romantic period: he had recognised in Schopenhauer's philosophy the idealistic metaphysics, and the remnants of Christianity it contained, which he could not but oppose.

In the new phase of his life that then began, Paul Rée, a philosopher who also frequented the Wagnerian circle, had great influence upon him. New ideas changed him from German idealism: Rée was acquainted with English philosophy, with empiricism, with materialism and the French moral philosophers. He admired Rée and learned much from him. He was especially impressed by his friend's naturalistic approach in moral philosophy, and by the idea that egoism is central for the explanation of moral behaviour.

A second creative period now began. Nietzsche started a crusade against his former ideals, against romanticism, and idealistic metaphysics, in fact against what he had hitherto considered to be the essence of German culture. He had read A. Lange's *Geschichte des Materialismus* (History of Materialism), and was enthusiastic. It had given him not only an excellent orientation in the history of philosophy, but for the first time he had become acquainted with empiricism and realistic positivism. In spite of this, it must be said, however, that he never became an adept of Anglo-Saxon empiricism—he remained a German after all—and that he rejected materialism, which he considered to be shallow. He preferred the Latin spirit to the German cultural climate, but he had no good word for the English philosophers, because in his opinion they have no spirit *at all*. During the time of his friendship with Rée he had come to know them fairly well, and he must have absorbed the nominalistic heritage that was present in the work of the empiricists, as was so clearly explained by J.R. Weinberg in his *Abstraction, Relation and Induction*, heritage that was an excellent antidote for idealism. Indeed, nominalists only accept the existence of what is particular and consider the general to be a mere product of the mind. In order to clear his philosophical path from metaphysics and idealism, he borrowed Ockham's razor, setting himself the task to eliminate those concepts that were only constructions of the mind to which corresponded no reality, a task he accomplished *con gusto*.

In this second period, he was at first inclined to value the positivism of his time as a weapon against the Christian, idealistic metaphysics of the philosophy of Hegel and his historicism. It also seemed a good remedy against the overvaluation and hypostasis of the subject, of consciousness and self-consciousness. Not only the metaphysics of Hegel had to be fought, but also Schopenhauer's idealism and the identification of the Will with *das Ding an sich*. Later on, he became critical of positivism. It was an offshoot of the Enlightenment, the Promethean movement that claimed to realise the well-being of mankind through "objective" knowledge of the world. He considered its aim as a conceit and a self-conceit: positivism asserted to be a search for objective truth, but in fact it constructed a rational picture of the world tailored to the measure of man. He came to recognise that it was at the opposite of his tragic conception of the world and of human beings. The world is not there for men, it shows no rational order and beauty, it does not correspond to their needs and wishes, and it has no ultimate goal. It must be said, that though he was not a positivist, but he was nevertheless constantly interested in the natural sciences of his time, a fact that is often ignored. He read books in different scientific domains, especially physics and chemistry, but also biology.[26]

Nietzsche, in his first work of the years of maturity, *Menschliches, Allzumenschliches* (Human, All Too Human), at first follows his friend Paul Rée's view of the genealogy of morality, but soon he rejects the latter's psychological approach that loosens morality from its foundation in a transcendent reality, but leaves it intact as a human universal. He questions morality on a more fundamental level: if it exists, what kind of reality is it; does it not pretend to be something it is not? Is it not for its inventors an instrument to gain power? He comes to recognise that this is indeed the case, and, as we shall see, in his philosophy he will replace it with values each individual has to forge for himself.

Nietzsche, whose bad health had forced him to leave the university, became progressively lonelier. He will soon lose Paul Rée as a friend, because of a stupid misunderstanding concerning Lou Salomé. Nietzsche had found pleasure in the company of this young Russian girl, who was not only beautiful, but also intelligent, and he hoped that he had found a pupil in her. Both men soon fell in love with her, but she did not return their feelings. There were further misunderstandings, and the situation was poisoned by the repeated interference of Nietzsche's sister Elisabeth in the affair.[27] Nietzsche, whose health was already poor, became severely ill. As on other occasions, this personal drama turned into a catharsis, and Nietzsche began his third and last creative period, often called the prophetic period, in which he wrote his most famous work, *Also sprach Zarathustra* (Thus Spoke Zarathustra).

His anthropology

Primacy of the individual

His fundamental intuition has been "*es gibt kein gleiches*," literally "the same does not exist," which means that nothing ever remains the same, and that never two elements of reality of whatever kind are identical. All is continuously changing, nothing is repeatable or identical, neither in space nor time. This position is typical for nominalism, which is generally identified with the elimination of the existence of eternal ideas, forms, essences, natures, substances, etc., but it is often forgotten that this is only the consequence drawn from the ontological principle that all that exists is particular and that the general has no existence of its own. By adopting this principle Nietzsche had rejected the basic conviction of Schopenhauer that we are individuations, i.e., material realisations, examples of eternal ideas. These are in their turn the expression of the general first principle, *das Ding an sich*, or the Will. However, *das Ding an sich* is an empty shell; a metaphysical fiction. Nietzsche has consequently defended the extreme thesis that the ontological principle must be the primacy of the individual, and he has applied it in all the domains of his thought. As it is so crucial, it can not be repeated enough: what is abstract and general is always the product of our mind, whereas all that exists, be it things or living beings, is concrete and particular.

Physical reality not to be grasped by the usual ontological categories.

For Nietzsche, like for Schopenhauer, man is part of nature. In consequence his conception of physical reality has a bearing upon his vision of man. He rejects materialism, which was popular in his time. In his opinion atoms must be theoretical entities. He denies that physical reality consists of indivisible elements, like the atomists claimed. Though he had been far from brilliant in mathematics at school, he had been fascinated by the mathematical concept of infinity. His intuition was that like between two mathematical points on a line there are infinitely many points, in each element of reality must be hidden an infinite reality, and if this is true, then no element is in principle indivisible. Ultimately nature is not matter, but energy.

In fact atoms are invented because the way we conceive of physical phenomena reflects the peculiarities of the Indo-German languages. In these languages the most common construction is the combination of a subject, a verb and a direct object. Therefore the language users cannot but suppose that to each action, to each production or working or movement, corresponds somebody or something that produces, works or moves. It is simply our way to conceptualise and understand reality to split it up and regiment it in facts, events, causes and consequences, because we need a reality divided into neat entities, ultimately in atoms, that are *constant and identical in their generality*. Were it not that we are tied to the logic and the grammar of our language we could easily invent other conventions.[28] In other words, the categories are not universal, a priori given, but invented by the members of a linguistic community. He states:

> The creative force that has invented the categories, was at the service of our needs, namely security, rapid comprehension on the base of tokens and sounds, means to abbreviate:—substance, subject, object, being, becoming, are not metaphysical truths.—The powerful were those who have turned the names of things into law and under the powerful it were the greatest artists of abstraction who have created the categories.[29]

Knowledge

Because Nietzsche accorded more importance to the direct human experience than to rationality, he did not share the common belief that human beings are superior to animals because of their capacity to think abstractly. A consequence of his general nominalistic conception is that no element of reality must be seen as an example of a kind, a genus, a species, but rather the other way round, the kind, the genus, the species are derived. They are constructions of the mind, necessary for a convenient representation of reality and therefore for survival, but still no more than representations and approximations. Also, these mental constructions are in no way objectively given, but are built in function of the particular interests of the human animal. If we were not the beings we are, but different animals, we would use different concepts. Knowledge for Nietzsche is not the limited *reine Vernunft* (pure reason) of Kant, but "*die grosse Vernunft*" (the great reason). His epistemology is not a search for the limits of knowledge based upon empirical data and rational categories, but his thesis is that knowledge consists of our lively experience and the grasping of features of reality that are of interest to us. It must not be true in a metaphysical sense, but it must serve the survival of individuals, i.e. Life.

Nietzsche's conception of knowledge resembles in a more than superficial way the epistemologies of the empirio-criticists R. Avenarius and E. Mach, who had a great influence upon the logical empiricists of the Vienna Circle. They considered knowledge to be a prolongation of common sense and everyday knowledge, corresponding to the biological needs of the species and necessary for survival. The conceptual scheme we use is therefore not given, but produced. This means that we can look at it critically and strip suspect concepts from their ontological content. In fact, Nietzsche did the same thing from a broader philosophical point of view. He adopted a meta-level perspective, and was suspicious of all the concepts he considered to be invented for the enhancement of the power of individuals of a species over their environment or the power of individuals in a certain society. Viewed from the meta-level these concepts were an encumbrance for sound thinking, and we must not be lead astray by them. He wrote for example:

> There is neither "mind," nor reason, nor thinking, nor consciousness, nor soul, nor will, nor truth: they are all fictions that are useless. What is given is not "subject and object," but a definite kind of animal, which only thrives under a certain relative *rightness*, before all *regularity* of its perceptions (in order to be able to capitalise its experience).
>
> Knowledge works as an *instrument* of power. Therefore it is evident that it grows with every increase in power. The sense of "knowledge" is here like with "good" or "beautiful," to be understood as strictly anthropocentric and biological. In order that a definite species is conserved and grows in its power, it must in its conception of reality capture so much that is predictable and constant that a scheme for its behaviour can be constructed from it. *The usefulness for survival—not* a kind of abstract/theoretical need not to be cheated—is the motive behind the development of the organs of knowledge...; they develop in such a way that their observations are enough for our survival. Put otherwise: the *measure* of the will to know depends on the measure of the growth of the *will of power* of the species: a species grasps reality *to become master of it, to put it at its service* and no more.[30]

Nature consisting of countless unique elements is "a necessary chaos," which means that, though we can discover similarities and regularities, it is not built according to a rational order. The laws we "discover" in nature are not ontological data, they are generalisations and abstractions, they are our creations and they are temporarily useful, but they do no justice to the reality they are derived from, which is much richer and in consequence cannot be grasped by universal propositions. Even the logic we use for establishing scientific truth is put into question: is it not a mere habit of thinking?

Reality is the reality we *experience* for Nietzsche; it shows an extraordinary diversity and a complexity that never can be rendered by the crude generalisations and approximations of science. It is from this point of view that he considers the glorious Newtonian system hailed by the great philosophers of the Enlightenment, Hume and Kant. It is a beautiful and rational abstract model of reality. In his opinion, the triumph of materialistic mechanicism has to be brought into a larger cultural perspective. As we have seen, matter is not the ultimate reality, but it is *energy*. More exactly, the unity of energy is a *quantum*, and the quanta reality is composed of are in innumerable relations with each other. What we see is the outcome, the result, of the combination of all these quanta or working forces that balance each other or are diminished or enhanced by each other in this network of interchanges.

The need we have to establish truths or even *the* truth is the consequence of a moral urge: we do not want to cheat and we do not want to be cheated. Though scientific truth has its benefits, in many cases, lies, dogmas, falsehoods are necessary for the prosperity of human beings. As Nietzsche stresses again and again, many cultural products, religion, metaphysics, art, must be considered from this point of view. Above all the belief that the world has a hidden sense, that it has a purpose, must be regarded as the expression of ineradicable *wishful thinking,* not as a truth, but as a useful fiction.

The same approach is chosen in ethics: morality must be considered as necessary tool for the survival of a society and for the establishment of the rule of one social group seeking to consolidate its power over other groups. In other words the Good has no foundation in the Truth as was believed by Socrates and Plato. Instead of coinciding with the Truth, it is a mere social necessity without foundation in a transcendent reality or in a rational model preceding its concrete realisation.

Subjects coincide with their bodies. There is no substance called "self," and we are not free.

Human beings are, like the rest of nature, composed of quanta of energy of which the interplay constitutes the unique combination of forces at work in our bodies. These bodies are the units of Nietzsche's anthropology, they are *objects that experience reality,* consciously or unconsciously, and from this follows that there is no strict distinction between body and mind. Neither is there a strict distinction between bodies and the reality they constantly observe and with which they interact. In other words, subject and object merge. And it must even be concluded that there is no strict difference between culture and nature, as the former is a product of the latter.

The main dogma of Christian culture, namely that human beings are essentially spirituality, soul, subjectivity, is denied and with them free will is thrown overboard. Hume, Schopenhauer and Nietzsche are great exceptions in the history of philosophy: they reject the existence of an "ego." However different their philosophies, all three have discarded the idea that we are inhabited from birth on by an unchanging entity, the "soul" or "self." In addition, Nietzsche, considering the many aspects of the problem and sometimes even repeating himself, denounces constantly the importance accorded to *consciousness and self-consciousness*. These central concepts of idealism are to a large degree biologically speaking superfluous. Consciousness and self-consciousness is in many cases the enemy of the force of life, and a decadent phenomenon.[31] In the idealistic conception man is a *historical being*: he has a historical mission, and must understand himself as an instrument for the realisation of a general aim. Historicism was fashionable in Nietzsche's days and he already warns against it in an early text, *Vom Nutzen und Nachteil der Historie für das Leben* (About the Usefulness and the Disadvantage of History for Life), one of the four *Unzeitgemässe Betrachtungen* (Untimely Meditations), in which he criticises the culture of his own time. An overdose of consciousness of the past can namely be harmful, and hamper the direct experience of life that is the source of creativity.

The fact that we can distinguish different periods in Nietzsche's life that are reflected in his philosophical work is often neglected. His credos changed over time and if his ideas are not put in the right context it is easy to attribute opinions to him that he only temporarily held. Moreover, as is well-known, he wrote apho risms, short remarks and reflections on different topics. This makes it even easier to distort his thought by selecting only those ideas that are grist on one's mill. Nevertheless, most of the themes we hitherto mentioned can be found in the writings of all the creative periods in his life; they constantly recur in one form or another. Those existentialists who have thought they could recognise in Nietzsche a forerunner, were certainly wrong. For one thing, what he meant by experience was not self-reflection. Existentialism is an offshoot of German idealism. Many of the ideas we can find in the work of Sartre and other existentialists stem from Hegel (the crucial distinction between *an sich* and *für sich* that Nietzsche explicitly rejected for example), the other main source is Husserl, another idealist. The experience of the self, to reflect on oneself and on one's own existence, is central in existential philosophy. It gives the human being access to experiences of a transcendent reality (Heidegger); it enables him to forge projects that shape his own life, to take responsibility for it, to relate to other persons, to exert one's free will (Sartre). Nietzsche, on the contrary identified himself with the unreflective vitality of his body and its instincts, with the Dionysian dimension of the individual, not with the cool Apollonian rationality. He is the first philosopher to value the instincts positively. Nevertheless, he never became a vitalist in the current sense, because he did not conceive of life as a *substance* passed on by living beings from generation to generation, but as a succession of individuals. It is another common flaw to see in him a representative of this movement.

Nietzsche has often thought about free will and considered its many aspects. In the first place he regards it as a psychological datum; human beings think they are free to decide what they are going to do. He uses the comparison of the silkworm finding his freedom in the spinning of its thread.[32] There are several reasons for this. In the first place, when we feel the force of life, our energy and strength working in ourselves, when we are angry, stubborn, when we hate or love, we have the impression that something in us, our "self" or "ego" is the cause of these feelings. We, humans, are so proud of our humanity that we want to be *causa sui*, our own cause. The half educated have the metaphysical need to be responsible for their actions and to discharge God, the World, History, Chance, Society from this responsibility.[33] We are used to a mechanical view of change in reality and not to think in terms of processes, but of cause and consequence. We have made separate things of them: for each "working" (consequence) there must be a "worker" (cause). If something happens in us, if we do something, there must be "something" that is responsible for it, our "ego" or self. Next to the psychological aspect of free will the theological aspect has to be considered. From old, the existence of human beings, their passions, their worries, their sorrows, their adventures, the wars they fight have been followed with interest by the Gods. Our doings have been to the Gods a theatrical festival, as is described in the tales of Homer. Jahweh too is a spectator watching his creatures, not only the history of individuals, but also of tribes and their wars and of his Chosen People. He even is "the inspector of men's heart and kidneys" judging their acts, their gestures, but also their intentions under diverse circumstances. The spectacle is only interesting if the outcome is unpredictable; therefore the actors must be free to choose their behaviour, and in the Christian conception, even to control their intentions, the reason why they do something, known to them by introspection. There is also a social aspect and a political aspect to the question. Once, evil deeds were not punished, but the debt made by harming others could be bought off, and this not necessarily by the perpetrator, but by his children for instance.

Later on individuals were judged for their acts, and made personally accountable. However, in order to be responsible, they had to act freely, not to be determined by the circumstances or by their specific personality, etc. The powerful judged the weak individuals, made them obey their rules.[34] At the core of the psychology that makes individuals feel responsible for their deeds is the belief that the cause of each act must be sought in their own consciousness, their inner self. In all times the priests have tried to gain power over simple men by making them aware of their appalling sinfulness, and they have threatened them with the most terrible punishments after death to make them obey the rules.

Finally free will has been invented in order to restrain our instincts. Nietzsche denounces Christianity as a religion that is the enemy of Life. The ethical values it proposes are meant to curb our vitality, to turn all that before was considered beautiful and strong into what is mean and loathsome. Christian morality is a kind of vampirism, sucking away the blood of the healthy individual. The Christian God is a synthesis of all that is contrary to Life. The concept of a transcendent reality, a world behind the world, has been invented to devaluate the only world there is, *this* world. The concepts "Mind" and "Eternal Soul" are invented in order to discredit the body, to create a qualm for the flesh. "Hail" has been put before "Health" in a kind of hysterics of Salvation. The concept of "Sin" together with the concept "Free Will" has been invented to tame the instincts. The ideal of the strong man who enjoys his passions, who follows his instincts is rejected and is replaced by the ideal of the altruistic individual, the person who is able of self-denial. This is pure decadence! Nietzsche asks "Have I not been understood?—Dionysus against the Crucified...."[35]

Schopenhauer denied that the world was divine, rational, justified. He saw Nature as cruel, a vale of tears, life as full of misery and suffering. There could be no hope, because it was the very essence of all that exists to be the expression of the Will. Those who reflect philosophically cannot but accept the world and themselves as a part of it for what they really are. Nietzsche recognised in Schopenhauer's attitude a profound pessimism and nihilism. The instincts and passions had to be suppressed as causes of disaster. He preached abstinence and that we had to be compassionate. These admonitions were remnants of Christianity. Nietzsche, on the contrary, rejected nihilism and pessimism. He loved life as it is, cruel, brutal, an ugly spectacle, but also delicious, marvellous. He has many times expressed the pleasure to exist, this ease and this satisfaction with life, with the earth, with his body, but perhaps never so perfectly as in "Midday" in *Also sprach Zarathustra* (Thus spoke Zarathustra).[36]

Anti-humanism

Though Nietzsche was not a system builder, his fundamental theses form a coherent totality. One of these is that God did not create the world, that the world is not a harmony. It is a "necessary chaos," i. e. the result of both chance and necessity and has no intrinsic sense. Indeed, if it would be completely necessary, its meaning could be found in a pre-ordained rationality, but if, though lawful, it is the result of chance coincidences, it is not justified by its own necessity. Another important position is that he denies that man is a reality apart and stresses the continuity between all forms of life. Already in *Menschliches, Allzumenschliches* (Human, All Too Human), and from then on again and again, he ridicules anthropocentrism. Perhaps the most beautiful early text, worth citing for its acidity and accuracy, expressing his anti-humanism is the following:

> *Man, the comedian of this world.*—There should be more spiritual creatures than men for totally savouring the humour of the fact that man considers himself to be the aim of the total reality of the world, and that mankind is only content with a world mission. Has a God created the world then he created man as the monkey of God, as a continuous occasion for him to cheer up in his all too long eternities. The music of the spheres around the earth would then be the ironic laughter of all the other creatures around man. The bored immortal tickles his favourite animal with sorrow, to find joy in the tragically proud gestures and explanations

> of his suffering, in general in the inventiveness of the vainest creature. Because, however, God invented man for fun, he had more spirit than man, and also enjoyed more the mind —....[37]

Nietzsche then continuous to ridicule the pretension of man to be incomparable and fantastic, to be unique in the world. But that we are unique is totally improbable. According to the astronomers, whose point of view is not tied to the earth, the drop of life on our planet is completely insignificant for the character of the immense ocean of birth and destruction of the Cosmos. Uncountable many celestial bodies offer similar conditions for the production of life as our planet Earth. Though they are many, they are only a handful in comparison to the vast majority that never have suffered from or are since a long time cured from this "living rash." Life on these celestial bodies has been, measured by the time of its existence, only a moment, a flaring up, followed by immense stretches of time after them. How could we then be the last end and aim of the world? It is clear from all this that many of Nietzsche's ideas were far in advance of those held in his time.

Evolutionism

Nietzsche was acquainted with evolutionism and the idea of continuity between man and the rest of nature, so strongly present in Schopenhauer's vision, is the complement of his idea that though nature has produced man, it is not there *for* man. He was aware that Schopenhauer could have been an evolutionist himself, but that his idealism excluded this option.[38] Nietzsche himself had first been an adept of Lamarckism, but when confronted with Darwin's theory, he became for a while his admirer, but nevertheless he remained cautious. Nietzsche came to know the theories of Darwin through his readings of authors commenting upon *The Origin of Species* and probably through a colleague, a biologist of the University of Basel, Ludwig Rütimeyer, who had met Darwin. There is to my knowledge no proof that he read *The Origin.* If he had read it he would have known that Darwin himself considered individual variability, not the species, which for Darwin is not an entity, but a sum of individuals, to be the motor of evolution. He would also have known that he did not see each step in the evolutionary process as a progress. Anyway his criticism is philosophical rather than scientific, and he is inclined to confound "is" and "ought." Because he was a lay in biology, it is fortunate that in his argumentation he does not use concrete biological examples, like so many philosophers have done with the most ridicule effect.

Nietzsche develops over the years a strong bias against English culture in general and English philosophy in particular, and even though he also often distances himself from his own German culture, he believes the latter is superior in many respects. This is reflected in the way he critically evaluates Darwin: "he is an Englishman after all." What disturbs Nietzsche is that his theory of evolution is Malthusian, i.e. that it is about the growth of populations compared to the growth of the amount of available food that is present. As we have seen, when considering nature, Nietzsche proposes as a general principle the existence of basic units of quanta of energy, forces, "struggling" with each other to gain supremacy over one another. When considering society, he did not refer for the explanation of social and political phenomena to the economy underlying them, but to the actions of individuals with very different qualities, and to a struggle for power between them. The threat existed that the vast majority of the weak and average individuals would grasp the power, while only the exceptional individuals, possessing "virtus," i.e., excelling by their qualities, could bring a creative change. Following Nietzsche, in Darwin's view variety in a population meant that not those individuals excelling most would be the fittest and survive, not the marvellous exceptions, the "bold and the beautiful" capable of taking risks fancied by Nietzsche. Those who, by chance being best adapted to the circumstances, would be able to eat most and to reproduce most, an idea that must have repelled the aristocrat in the philosopher. It is my conviction that he knew that Darwin was fundamentally right, but could not believe that the Principle of Life was a vulgar matter of purely material benefit. In his view, it had to be related to gain of power by mental qualities, i.e. a quick grasp of situations and of similarity in the diversity, and furthermore cunning and courage.[39]

He is also wrong in believing that Darwin assumed that the tree of life shows the evolution of ever "better" species, that evolution means progress of the species. However, his criticism was in his later years more directed against the Darwinians and their interpretations than against Darwin. In one of the writings that remained unpublished during his lifetime, and which can be considered to contain the conclusion he came to, we read:

> *Anti-Darwin.*—What surprises me most in an overview of the great destinies of man, is to have always before my eyes the contrary of what today Darwin with his school sees or *wants* to see: the selection in favour of the stronger, those who are better off, and the progress of the species. Just the opposite presents itself to us: the erasing of the lucky cases, of those types that got higher, the middle cases becoming master, even the *less than middle* types. In case one cannot show us the reason why man is the exception under the animals, I am inclined to the prejudice that the school of Darwin has been mistaken. This Will to Power in which I have recognised the last fundament and character of all change, gives us the means to explain why it is just the selection in favour of the exceptions and the lucky cases that does not take place: the strongest and the happiest are weak, when they have against them the organised instincts of the herd, the fearfulness of the weak, the greater numbers. The general aspect of the world of values I posit shows that not the lucky cases, not the selection/types, have the supremacy in the highest values hovering above

> mankind: rather the *decadent* types,—perhaps there is nothing more interesting in the world than this *unwanted* spectacle....[40]

It is odd, he says, that one has always to *prove* the strong individuals against the feeble individuals, the children of fortune against the failures, the healthy individuals against the genetically defective. This is the morality of Christianity, of Schopenhauer, of Buddhism: the denial of the Will to Live. I hate Christianity for this, he says, because it is a justification of this dreadful, grim world, because with sublime words and gestures it gives it the appearance of justice, virtue and divinity, and last but not least of morality. How is it possible, he asks, that philosophers and men of science alike admire this theory of the *reversed* struggle for life: a species is powerful if those individuals that do not express the force of life climb to the top. The mistake of Darwinism is clear, only how can one be blind enough not to see it?

Indeed, in the evolution of the species we cannot see a *progress;* they are simply *levels.* It has not been proven that higher species evolve from lower species; on the contrary those who are in the majority have an advantage. It cannot be explained how a random change in its genetic make-up can give a species a benefit over long periods. In short, nothing proves that the fittest are also the best possible. They simply have time on their side, they live longer and reproduce more than those strong individuals that excel in all respects, that are most exposed to danger and have a short existence.[41] It must be said, that Nietzsche sensed some of the key problems present in the Darwinism of his time. Soon the original theory was interpreted in such a way that the sting present in it was taken out. It was made inoffensive by turning it into a progressive history, from the simplest organisms to more and more complex organisms, from mindless organisms to creatures with a certain intelligence, and then to the most marvellous species: rational Man. He was right to reject this representation of the tree of life, which would remain common even in the twentieth century. In addition, he was also right to think that the basic unit in the history of life could only be the individual, and more especially the exceptional individual, not the species, which would become central in certain forms of Darwinism, though it was not in the original theory.

It is unclear whether Nietzsche supposed that Darwinism necessarily implies a teleological of "utility," which depends upon the living conditions of a population. In his opinion, no organ is formed because it is useful, organs are formed by the "internal" conception, but it is certain that he had questions about the concept form creating violence (=force) that profits from and uses the external circumstances, rather than that the external forces give the impulse for evolution. (Today a number of biologists, most of them morphologists studying the development of organisms, present theories that go in the same direction: matter, and especially organic matter contains self-organising principles and many features can be explained without seeing them as adaptations).[42] In conclusion we can say that from early on he became an evolutionist, but not a

Darwinist, because on philosophical grounds he could not accept that the survival of the fittest could be the principle of evolution. Mere chance could not be creative; where men were concerned it had to be based on "*virtus*," personal excellence and merit. This proves that indeed, he confounded in this domain the "is" of science and the "ought" of his philosophy and that he preferred the "ought" to the "is."

The reversal of all values

Another major novelty is his "amorality" and his "*Umwertung aller Werte*," the reversal of all values. In his opinion Schopenhauer's error was to rely on a combination of Buddhism and remnants of Christianity, and being profoundly pessimistic, to turn away from life, to preach asceticism and compassion, to prefer non-being before being. Nevertheless, he liked his "almost English" clarity, his honesty that found its expression in admitting that reality is not a rational order, that we do not live in the best of all worlds, but rather in the worst possible world. He also admired him for his unrelenting denial of the existence of the free will, free will that he himself, like the soul and the self, tirelessly denounces as a fiction.[43] What attracted Nietzsche above all was his consequent atheism. Central in his own thought is the fact that God is dead, Christianity is declining: "The greatest new event—that "God is dead," that the belief in the Christian God has become incredible—begins to throw its shadow upon Europe...."[44] All that was founded upon the Christian faith has now to be questioned. Without the belief in a transcendent reality that is its origin and foundation this world has no intrinsic meaning. It is clear that without a reality beyond reality no ethical system can be built, because for this purpose at least a transcendent first principle is required as a guarantee of the validity of the universal values proposed. Not even morality has any ground, except that it benefits to the priests and the ruling class, but nevertheless it probably is an indispensable element for the survival of society. For the philosopher, however, traditional values are like heavy stones that have to be turned over, examined, cleared away and replaced by totally new ones, often by their contraries. Nietzsche freely produces new values for himself, derives them from his own experiences as an individual that is a body and soul in one; they are his most intimate personal creation. Others can share them if they are his "*homoioi*," individuals who are like-minded, and therefore of the same disposition, or they can reject them if they want to keep to the traditional values. This is the meaning of the motto of *Also sprach Zarathustra*: "a book for everybody and no one." He thus combines amorality, i. e. the rejection of morality generally valid in a society, with an extreme individualism that bears a certain resemblance to that of Max Stirner, whose work *Der Einzige und sein Eigentum* (The Ego and its Own) he had read and even recommended to a friend.[45]

Nietzsche lived in the second halve of the nineteenth century, and his work is a protest against Victorian morality, against a bourgeois mentality, against the "*Bildung*" (cultural education) of his time It is a reaction against the Christian vision of man as well as against the socialist vision. In the last period of his life, sometimes called the prophetic period, he wrote *Also sprach Zarathustra*. Here he identifies himself with the oriental prophet and speaks through his mouth. His prophecy is that either "*Last Man*" will come, the true representative of his contemporaries, showing all their faults, their vulgarity, their meanness, their cowardice, their lack of pride, or "*Übermensch,*" (Upper-Man or Over-Man), the successor of man will evolve from man and replace our species that will become extinct. This is certainly a daring idea: nobody had ever seriously taken into consideration that our species was still evolving and would possibly one day disappear and be replaced by a totally new kind of being, let alone see this as a hopeful event. The alternative, the continuation of the species, and then the arrival of Last Man will be the fate of mankind if strong exceptional individuals do not bring about a profound change. This Last Man cherishes the Christian values: compassion, love of the neighbour, meekness, humility, *and* he cherishes the values of socialism that was in the lift at the time, namely equality and material well-being. Nietzsche loathed the plutocracy, feared the replacement of the traditional agrarian society by the industrial society that was in the make. He equally feared the rise of socialism, which would bring a general massification with it that was a threat to the individualism he dreamed of. He probably partly forged under the influence of Jacob Burckhardt, whom he knew well during his academic career, his Renaissance ideal of a society ruled by creative and strong men. Summarising these ideas in a prophetic call and warning, Zarathustra thus addresses the masses:

> When Zarathustra came in the following town, which lies near the woods, he found there many people gathered on the market, because it had been promised that they would see a tightrope walker. And Zarathustra spoke thus to the people: I learn you Upper-Man. Man is what should be overcome. What have you done to surmount him?
>
> All beings up till now created something that was above them: and you want to be the ebb of this great flood and want more readily go back to the animal than to overcome man?
>
> What is the ape for man? A laugh or a painful shame. And just this man will be for Upper-Man: a laugh or a painful shame. You have gone the way from worm to man, and many in you is still worm. Once you were apes, and even now man is more ape than any ape.
>
> He who is the wisest of you all is only a split and a mix of a plant and a ghost. But do I say you must become a ghost or a plant?

Look, I teach you Upper-Man!

Upper-Man is the meaning of the Earth!

I beseech you, my brothers, remain faithful to the Earth and don't believe those who speak to you about supra-terrestrial hopes! They are poisoners, whether they know it or not.

Those who despise Life are dying people and they themselves poisoned, tired with the Earth: so they can better disappear!

Once the resentment of God was the great resentment, but God died, and with that the resentful died. To resent the Earth is now the most terrible and to think the bowels of the inscrutable are higher than the meaning of the Earth!

Once the soul looked upon the body with loathing: and then this loathing was the highest—it wanted the body skinny, horrible, starved. So it thought to escape from it and from the Earth.

Oh, this soul was itself still skinny, horrible and starved: and cruelty was the lust of this soul!

But you too, my brothers, say to me: what news brings your body from your soul? Is your soul not poverty and filth and a pitiful delight?

Truly, a filthy river is man. One has to be a sea all right to be able to receive a filthy river without becoming impure.

Look, I teach you Upper-Man: he is this sea, in him a great loathing can go under....[46]

When the tightrope walker is in the middle of the rope a jester cries out and jumps over him, bringing him out of balance and he falls in the midst of the frightened people on the market place. It now has become clear how Nietzsche used the metaphor of the Übermensch to signify that man must not see himself as an ultimate reality, but as a bridge to something that did not yet exist. He is a cord stretched between what was and what will be; like the tightrope walker he is a transition and his own destruction. But the danger remains that Last Man will win. He is in fact the domesticated form of man, fed and pampered by the new industrial society, man who has lost his wildness.

Nietzsche's Upper-Man, has been misunderstood in various ways, as a superior idealistic type, partly saint, partly genius, as a Darwinian well-adapted type, or even as a hero, though Nietzsche often expressed his scorn for the cult of heroes. His contemporaries did not understand that Upper-Man simply could not be subsumed under one of the usual concepts of their culture, but that he is on the contrary the destroyer of all the traditional values.[47] What we do know about this Nietzschean thought-experiment is that this new being will be "beyond good and evil," strong and self-disciplined instead of weak, that he will accept his instincts and passions, love the Earth and his own body. However, it is possible that Last Man will overrule the courageous exceptions and that he will thrive and become "numerous like the flee-beetle."

Zarathustra is aware that people won't listen to him. He wants to make them listen by putting to the trial their pride and speak of what is most loathsome, namely Last Man. He warns them that it is yet time to prepare the coming of Upper-Man, but that time will run out and that soon they will be unable to let him grow out of them and let him come into being. In that case the triumph of Last Man is near. Last Man is the most despicable of all men as he cannot despise himself.

> See! I'll show you *Last Man*.
>
> "What is love? What is creation? What is passion? What is star?"—asks Last Man and blinks.
>
> The Earth then has become small, and on it hops the Last Man, who makes everything small. His species is inexterminable like the flea-beetle; Last Man lives longest.
>
> "We have invented happiness"—say the Last Men and they blink.
>
> They have left those places where it was hard to live, for one needs warmth. One loves the neighbour and rubs oneself against him: for one needs warmth.
>
> To become ill and to be suspicious they hold for a sin: one walks carefully. A fool who still stumbles over stones or men!
>
> A bit of poison from time to time: it gives pleasant dreams. And much poison at the end, for a pleasant death.
>
> One still works, because work is an occupation. But one sees to it that the occupation does not exhaust.
>
> One does no longer become rich or poor: both are too troublesome. Who does still want to rule? Who still wants to obey? Both are too troublesome.
>
> No shepherd and *one* flock! Everybody wants the same, everybody is equal: who feels otherwise goes of his free will in the asylum.
>
> In the past everybody was wrong—say the smartest and wink.

> One is clever and knows everything that has happened: thus there is no end to mockery. One still quarrels, but soon follows reconciliation—otherwise it ruins the stomach.
>
> One has one's little lust for the day and one's little lust for the night: but one honours health. "We have invented happiness"—say the Last Men and they wink.
>
> When he had thus spoken the people interrupted him shouting "give us this Last Man, oh Zarathustra, and we give you your Upper-Man!"[48]

The Eternal Returning and "Amor Fati"

His contemporaries, as could be expected, did indeed not understand Nietzsche. They were not prepared after nearly two thousand years of Christianity to listen to a prophet preaching the reverse of the values that had over the centuries become a second nature to them. The love of what is earthly had been for so long sinful, and what really counted was the reward for a pious life. Here now was a man who told them that the only salvation lies in this world, the immanent world, that the transcendent reality, God, is an illusion. Nietzsche had one day, wandering about in Engadin, in Switzerland, an idea that struck him like thunderbolt. I mentioned already that he was fascinated with the idea of infinity. Suddenly he realised that if, like the Greeks thought, time is *eternal,* each moment with all its particularities will once return, not one time but infinitely many times.[49] This is a frightening idea, but, if time is indeed eternal, logically inescapable. He accepted this eternal returning: the strong individual does not moan and complain about his existence, about the hardships he has to endure in this vale of tears, but accepts the chaotic and senseless necessity of reality, he accepts that he is part of it and loves his fate: "amor fati."

Reception

Nietzsche's conception of man and his relation to the world was revolutionary, but what has been the impact of his ideas? The *interest* for Nietzsche's work in the previous century has been immense, but seldom were his ideas *understood* for what they were. Alfredo Guzzoni, the editor of a work on the reception of Nietzsche's work puts it thus:

> It is certain that Nietzsche's influence upon the philosophy of the twentieth century—compared to Kant or Hegel in their time and later—has been limited: his philosophy has neither in a lasting way put into jeopardy the existing view about man and the world, nor built the fundament for a new one. In spite of the variety of impulses, which his

> philosophy gave, it has in most cases been rejected or been integrated in an already fixed conception and interpreted in this perspective.[50]

I expressed a similar opinion in my unpublished doctoral thesis *Nietzsche's Influence* (1968), where I have shown that, though many European intellectuals were infatuated with Nietzsche's iconoclastic fury, and though in their own work they were certainly inspired by his ideas, they did not adopt them under their original form. These artists and philosophers had their own agenda. Nietzsche had admirers, but no influential pupils.

The Nazi philosopher Baeumler has used him for the ideology of the Third Reich. In the Thirties different authors have given a metaphysical interpretation, in accordance with their own views: Karl Jaspers, Martin Heidegger, e.g. An exception is Karl Löwith, who tried to understand the real historical meaning of Nietzsche's philosophy. This allowed him to grasp the core of Nietzsche's work: it is the attempt to restore to the world its primitive innocence by eliminating the Christian God who is the creator of the world and reigns over it and over all its creatures. Löwith considers this a triumph of the scientific mentality over the religious world view. In short:

> with the elimination of a conscious divine will, divine aims and a moral world order the world shows itself again, as it was originally: beyond good and evil, as an "innocence of becoming," including man, for whom nobody bears the responsibility, nor a God, nor he himself.[51]

Thus, according to Löwith, Nietzsche has freed man of sin and of a world mission. Finally after the Second World War the interest for Nietzsche was aroused mainly outside Germany and his work or aspects of it were analysed from different perspectives. Nevertheless, nobody changed fundamentally his mind about how to understand the human phenomenon. The Anglo-Saxons had generally no interest for a specific philosophy of man, and on the continent an idealistic anthropology prevailed as in the past.

Notes

1. Cf. W. Abendroth, Schopenhauer, Rohwolt, Hamburg, 1967, 37.
2. R. Safranski, Schopenhauer und die wilden Jahre der Philosophie, Carl Hanser Verlag, München,1987, 317.
3. Schopenhauer. Werke, Wissenschaftliche Buchgesellschaft, Darmstadt, 1982, II, 618.
4. Ibidem, II, 741.
5. Safranski, Schopenhauer, 317 and Schopenhauer. Werke, I, 157.
6. Safranski, Schopenhauer, 333.
7. Schopenhauer. Werke, III, 679 and following.
8. Ibidem, I, 549.
9. Ibidem, I, 657–658; II, 225.

10. Ibidem, I, 208–210.
11. Ibidem, I, 212–213.
12. Ibidem, I, 210–212
13. Ibidem, I, 215
14. Ibidem, III, 361–363
15. Ibidem, III, 364–365
16. Ibidem, III, 365–366
17. Ibidem, I, 218–219
18. Ibidem, I, 54 and 257.
19. Ibidem, I, 56
20. Ibidem, I, 507–508
21. Ibidem, V, 441
22. K. Schlechta, Nietzsche Chronik. Daten zu Leben, und Werk, Hanser Verlag, München, 1975, 24: the end of October/or November 1865.
23. Cf. Schlechta, Nietzsche Werke, 3 vol., Hanser, München, 1956, Band III, 961, 962, 963, 970, etc.
24. Curt Paul Janz, Friedrich Nietzsche Biographie, Hanser, München, 1978, 255–256 and passim.
25. Cf. Schlechta, Nietzsche Chronik, Hanser, München/Wien, 1975.
26. Cf. M. Gosselin,"Lou Salomé en Friedrich Nietzsche," Dialoog, nr. 3–4, 243–260
27. Cf. Schlechta, Nietzsche. Werke, II, 583§20; II, 600§34; III, 909; III, 914.
28. Ibidem, III, 914.
29. Ibidem, III, 751.
30. The examples of this tenet in his work are too numerous to be cited in ex tenso. Cf. Karl Schlechta, Nietzsche Index, Hanser, München, 1965. A collection of them could be: II 44, §11,II 220–222,§ 354,II, 226,§ 375, II 226 (against Descartes 's ego, reflective thought), III 711, III 751, III 682–683.
31. Schlechta, Nietzsche Werke, I, 877, § 9.
32. Ibidem, II, 585, § 21.
33. Ibidem, II, 976, § 7.
34. Ibidem, II, 1159, § 8 &9.
35. Ibidem, II, 512 and ff.
36. Ibidem, I, 879–880, §14.
37. Ibidem, II,104, § 99
38. Ibidem, II, 999, § 14.
39. Ibidem, III, 748–749.
40. Ibidem.
41. Ibidem, III, 889–890. Cf. Brian Goodwin, How the Leopard Changed its Spots. The Evolution of Complexity, Phoenix, London, 1979.
42. Ibidem, I, 513, §107, I, 877, § 9 & §10, I, 887, § 23, I, 1098, §126, §128, II, §233, §360, III, 536–537, III, 612 (immortal soul), III, 913.
43. Ibidem, II, 205, § 434.
44. Cf. Nietzsche Chronik, 49.
45. Schlechta Nietzsche. Werke, II, 279, § 3.
46. Ibidem, II, 285, § 6.
47. Ibidem, II, 1101, § 1.
48. Ibidem, II, 283–285, § 5.

49. Cf. also K. Löwith, Nietzsches Philosophie der ewigen Wiederkehr des Gleichen, Kohlhammer, Stuttgart, 1956, specialiter 192–193.

50. A. Guzzoni, 90 Jahre Nietzsche Rezeption, Hain, Königstein, 1979, X.

51. Ibidem, K. Löwith, "Nietzsches Versuch der Wiedergewinung der Welt," 101.

Chapter 9. The twentieth century: more of the same

German philosophical anthropology

Nietzsche became famous during his lifetime, but was not understood. It was not clear to the European intelligentsia that he had not only broken with the ideology of nineteenth-century culture, but that he had revolted against the creeds of a civilisation that had lasted for more than two thousand years. After him the general vision of man remained to a large degree idealistic: man is a subject, a "self," an "I" in a body to which it is tied. Man's spirituality makes him a special reality, different from the other living beings and of course from vile dead matter. His self-consciousness guarantees his capacity to be free and also to relate to a transcendent reality, Being, or perhaps God. Animals *are* their bodies, can be defined by their bodies, and *if* they are accorded feelings and thoughts it is their *bodies* that feel and think. Men, not being their bodies, but subjects able to reflect upon themselves are spiritual. This is where, not the gap, but the abyss that exists between science and the rest of Western culture, religion, metaphysics, ethics, becomes obvious. Nietzsche's views were well in accordance with the science of his time and in some respects even with the science that had yet to come and this explains that they were *unzeitgemäss*, i.e., not in accordance with the general mentality of the second half of the nineteenth century. Karl Löwith is one of the very few authors who has recognised the discrepancy between Nietzsche's philosophy and the spirit of his age and who assessed his thought rightly as "a triumph of scientific mentality over religious spirituality."[1] Most Westerners ignore the scientific viewpoint or at best take a defensive position; they are subjugated by the general ideology that is incongruent with it. They claim, and rightly so, that science cannot and does not want to give answers to such questions like "what is the totality of reality?," "what is Being?," "what is ultimately our human nature?," "what is the sense, the meaning of our existence?," "what is the sense, the meaning of the world?," "what is good, what is evil?." What they forget is that in order to be able to answer these questions it would be better to be well-informed; the scientist cannot furnish the answers to these metaphysical questions, but, if they must be asked, he can give the philosophers an enlightening insight in the state of affairs in reality that matters for reaching a conclusion.

In the Anglo-Saxon countries generally speaking the main concern of philosophers in the twentieth century was epistemology, not anthropology. On the continent matters were different: here philosophy did not concentrate upon the foundations of knowledge or upon an analysis of formal and/or daily language, but upon questions directly related with the ontological status of man.

At the dawn of the previous century, H. Bergson created a new metaphysics that started from the theory of evolution, but derived from this theory the idea that Life is essentially a creative force. E. Husserl's phenomenology was a source of inspiration for a group of philosophers interested in developing a new philosophical anthropology as an answer to the scientific theories that were an obvious threat to the traditional metaphysical view of man and of the world.

The German philosophical anthropologists have in common that they reacted to the formidable progress in science and more especially in biology by rejecting a naturalistic approach of man and nature. They did not so much want to oppose scientific theories, but to show that from a metaphysical point of view scientific data are only raw material. What is of importance for a human individual is not merely to *know* about himself, but to *understand* his being, his essence, his place in the totality of reality, his task in this world, the sense and justification of his existence. The ultimate goal of science was a complete and unified description of all the domains of nature, while what they wanted was to show that reality consists of qualitatively different layers of being, between which no transition exists, and that cannot be reduced to each other, because each new layer shows characteristics that cannot be translated into characteristics of a previous layer. For these philosophers the new tendency to bridge the gap between the humanities and the natural sciences could not succeed, because it was artificial, being the result of epistemological considerations that were not rooted in a "first philosophy" clarifying the basic ontological categories, the *Seinsschichten* or layers of being. In their opinion, science cannot understand itself without first establishing the true nature of the human beings who create it.

Helmuth Plessner

Together with Max Scheler, Helmuth Plessner founded philosophical anthropology. Plessner did not reject the biological theories of his time, but on the contrary believed he could find in them the basic evidence he needed for his anthropology as a separate philosophical discipline. He returns to the ancient Greek concept of nature, and does not discard knowledge of what is material from philosophical speculation. He opposes living matter to dead matter and distinguishes in the domain of living beings three ontological types according to their respective relationship with their environment. First there is the realm of plants; their form of organisation is "open," they depend directly on their environment and form a continuum with it. Another realm is that of animals that are "closed" or "centric" organisms; they are centred upon themselves in an environment they can perceive and to which they can react, possessing a central nervous system and sense organs, and being able to move. Finally man is a class apart, because in addition to his animal organisation, he is able to distance himself from himself, to be "eccentric." He is body and mind in one, and by his self-reflexivity he is to a certain degree able to determine what he will be, to create himself. The reason Plessner gives for the unique ontological status of man is biological: man is unique, because he has gradually lost his instincts and is not bound to follow a rigid pattern in his behaviour. This is what Plessner calls "plasticity," an important concept in philosophical anthropology.

Arnold Gehlen

Arnold Gehlen was Plessner's pupil, and it is not always clear what he did and what he did not borrow from his master. He considers man to be logically speaking imperfect, because he lacks the tools other animals possess to cope with their environment and because his instincts are weak. This constitutes his plasticity and can be explained by a special biological quality, "neoteny," which means that he retains for a long time the characteristics he had as newborn baby. This goes together with a prolonged state of dependency on his parents and a long period during which he can learn instead of having to rely on his inborn behaviour. Because he is ill-adapted to his environment, he must compensate for this in order to survive by creating artefacts and an artificial environment. Man is primarily an actor who continually transforms nature. Gehlen characterises him as *mangelhaftes Wesen*, an imperfect being missing something. However, this is not perceived as being negative, because this forces him to make up for his weakness and vulnerability and to transform nature under its raw form. It leads him to create culture consisting of "institutions" in a very broad sense, comprising the making of tools, being able to communicate inter-subjectively by inventing symbolic language, creating art, and most important inventing different forms of social organisation.

Nicolai Hartmann

In the beginning of his philosophical career Nicolai Hartmann belonged to the Marburger Schule, whose representatives built further on Kant's transcendental deduction. However, went his own way. He had always been interested by reality itself, rather than by the way we conceptualise it subjectively: the latter must be derived from the former and not vice versa. This is the origin of his version of the theory of the different layers of being. He returns to a kind of realistically interpreted Aristotelian ontology: he tries to grasp reality as it directly appears to us, in an act that points beyond itself and transcends itself, revealing reality as it is independent from us and before all knowledge. He distinguishes between the inorganic layer of being, the organic layer, the psychical (emotional) layer and the spiritual (intellectual) layer. By stressing the ontological distinction between these two last domains he reacts against the "psychologism" of his time, which reduces mental entities to psychological data, denying them an independent existence.[2] Each layer contains a specific "*novum*," something radically new that didn't exist before and that suddenly "emerges" from what existed already and that cannot be analytically derived from its parts.

He concentrates on problematic metaphysical questions that can never be totally solved, such as the fundamental forms of human being, existence, life, consciousness, mind, freedom. He tries to understand the ambiguous relationship between the human soul and its acts on the one hand and "*der Geist*," Spirit, on the other. Though many of these questions remain problematic, it is not a good policy to found this metaphysical reality he discovers, in God. Therefore, he adopts the postulate of atheism.

It is only by trying to solve these problems that we can discover our true humanity. Ideally we should start from spirituality, but we are unable to grasp directly what it truly is. Starting from our unique characteristics also offers only a partial solution. In his own words:

> If one would in seriousness define Spirit by referring to "man," then one should already before this know what the content of man is. But do we know that? One will say, to a certain limit, of course we do. But what are the limits inside which we know this? They are the limits posed to our knowledge of Spirit, because we do not mean man as body, or even as psychical being. We reach, when we want to determine what is specifically human without making mistakes for those characteristics like knowledge, being purposeful, understanding the sense of things, freedom, ethos, acting creatively, etc.—only for fragments of his spiritual being. We can only truly define man on the reverse, by our fragmentary knowledge of Spirit. This is instructive. This order in determination clearly cannot be reversed.
>
> One must not make the mistake that we can read the characteristics of Spirit in their turn from man. This has nothing to do with the determination of the concept. Surely man and his world is the field of the experience we have of spiritual being—but not in such a way that we gain first a concept of man from which we could then borrow definitions. Not from the concept of man, but from man himself we read the characteristics of the mind. And up to the point to which through this a concept of Spirit is formed, what is specifically human can conceptually be grasped.[3]

Max Scheler

Max Scheler is, like H. Plessner, a pupil of Husserl and, being a man of broad interests, he applies phenomenology in different domains, such as ethics, the philosophy of culture and of religion. Yet, his method differs in part from the "eidetic" reduction of Husserl: the search for the essence of things is not an intellectual process, but rather "a loving participation of a person in what is essential in the phenomena."[4] Nevertheless, he follows Husserl, when he declares that the ultimate form of knowledge is not the scientific investigation of concrete facts, the gathering of empirical data, the examination of their causal connections and of the laws of nature, but the search for the knowledge of the essences of the objects of our consciousness. The task of a philosophical anthropologist is to show how, from the fundamental way of being of Man, his *anthropina* or essential and unique characteristics, his achievements and his works can be deduced. What is needed is the clarification of the nature of the relation of man to the foundation of Being.[5] Three elements are determining in Scheler's thought: philosophy of life, German idealism and Christianity.[6]

Like Plessner he distinguishes between four layers of reality; dead matter, living matter, plants and animals. In fact time and again a *scala naturae* is presupposed. Reality is governed by two principles: Life and Spirit. The Principle of Life is a *dark urge*; living beings are aware of the reality outside them that opposes itself to them, limits them in their functioning. The primitive force present in them enables them to cope with the resistance exerted by their environment. This force, this urge that is at work, is an irrational drive more basic than any form of intelligence and it is present in all living beings. The first form of it we encounter in nature makes the survival of plants possible and plays a role kindred to that of the vegetative soul posited by Aristotle. Plants show external signs of their state, which is poor or healthy, with their leaves drooping or full of moisture, but they cannot communicate. This primitive force, however, is, though irrational, not dull, but on the contrary even here it is creative, as can be seen from the incredible diversity of forms and colours of plants. Because animals can move about, their needs are different, and in addition to the *Gefühlsdrang*, as Scheler calls the dynamic principle that is at work in them, there is the urge caused by their sensitivity; they are capable of experiencing the different features of their surroundings. They transmit the result of their perceptions to a centre that gathers the information and makes them alert to their situation and influences their behaviour. (As was normal in his time, he assumes wrongly that there is such a centre of coordination in our brain). Memory, the faculty to be aware of past situations, of precedents, is a condition for the capacity to learn that is present even in the most primitive animals. The instincts of animals, the compulsion to feed, to grow and reproduce, are all expressions of the dark energy at work in them.

The *instinct* is a further development of this primitive force; the behaviour it commands is meaningful, because it serves an end that is genetically determined and specific for the species. It proceeds following a "rhythmic," unchanging course and therefore in its rigidity cannot be adapted to novel situations. The next step in the development is only possible when animals become more individual, more independent of the instincts typical of their respective species and from then on the conditions are fulfilled for *associative memory*. This type of memory depends upon learning by repeated trial and error. The successes and failures are reported to the centre processing the information, and thus the individual can adapt to situations that are atypical of the species, but typical of its own conditions. *Practical intelligence* does not rely upon a repetitive learning process, but it is based upon insight.

Here we arrive at the crucial moment in his speculative representation of the hierarchy in the succession of species, where Scheler must draw the line between Man and the animals. Scheler does not deny that man has evolved from animals; on the contrary, he concedes that higher animals, and certainly the apes, possess practical intelligence. But if Man was merely "homo faber" (and as such he is represented by science), he would be only gradually different from the animals. Man, however, is not merely the expression of the creative Principle of Life, but he is related to the realm of the Spiritual. Scheler elaborates upon the problem of the ontological status of the psychical and the spiritual. Here, he says, the roads of those reflecting upon Man part: on the one hand some reserve intelligence and free will for Man, where he sees no essential difference; on the other hand, the evolutionists, whether Lamarckists or Darwinists, reject, like Darwin, G. Schwalbe and W. Köhler, an ultimate difference between men and animals. The latter group believes animals can be intelligent; they do not accept any metaphysical Being, and in consequence no metaphysics of Man: there is no specific relation of Man to the ground of Being. For them there is only a gradual distinction between Edison and an intelligent chimpanzee.[7] In Scheler's opinion, this is not to the point. Man does not earn his special place in reality by his intelligence or his free will, but owes it to a special capacity. It is by speculation that man can approach "*der Geist*," Spirit, and this capacity is not related to the Principle of Life, but a totally new dimension. Now we can see how the motives come together: Life is the more primitive element, Spirit is a late phenomenon, weaker than Life, more vulnerable, because connected to that most special ability of man, abstraction that is the basis for speculative thought. This is an original point of view, which, as he will stress, is contrary to Western tradition.

We must halt here for a moment and reflect upon the relationship in this type of thought, not between science and religion, but between science and metaphysics. Max Scheler does not oppose the findings of science, though he is most of the time critical of them. He accepts that man has evolved from other animals, and he accepts that higher animals are capable of thought and even that they are intelligent. He denies that man was a necessary outcome of the evolutionary process. Biological data are merely useful background information when he establishes what the different layers of being are, but it remains accessory for his aim, the search for the essence of Man. Science has nothing to say about this, it is indifferent to real philosophical thought; it has chased metaphysics away, banned it. He has the same complaint as Hegel and Heidegger: science has become quantitative, calculating and technical. A metaphysics of man, however, requires an attitude that engages the person of the philosopher in his totality; it is not the cold intellectuality of the scientist that is needed, but speculative intuition. He declares that the new principle that comes into being with man does not grow out of the evolution of life. What makes men human is not a degree of life, and it is not only the degree of what is psychical in him, but it is a principle outside life, that is even contrary to all forms of life, and even to man. If one would like humanity to be founded in something it must be in the supreme ground of all things of which life is a manifestation.[8]

What explains the special position of man in nature? It is with man that this totally new kind of reality comes into existence to the extent that Man is really a *person.* In order to explain what it means to be a person, rather than merely a living being, Scheler refers to a concept that is central in philosophical anthropology, namely the human capacity to *change* reality. Man's *Weltoffenheit*, his ability to adapt to his surroundings and to transform them, distinguishes him from the mere animals that are on the contrary completely dependent upon their environment in the way they function. The first *anthropinum* mentioned is this *being open to the world.* A spiritual being is no longer bound by his instincts and his environment. He knows about the mechanisms with which he can react to the world and resist it; he can turn them into objects of his thought and know that they are as they are. Thus he is no longer limited to undergo passively his own instincts and the data of his pre-programmed sensorial system. Spirit is to turn reality outside ourselves and ourselves as part of it into an object, and only he is the carrier of spirit who has the reverse relation to the world of that of the animals.[9]

Human beings are not compelled to stick to the "this and that" level, to the "being thus" of what they encounter, but they are motivated by representational and contemplative complexes, independent of their physiological states and of their instinctive impulses. They are capable of generalisation and abstraction. This is the basis for their free choice to refrain from following their instincts, for their capacity for abstinence and for grasping the essence of the phenomena that present themselves to them.[10]

A second *anthropinum* is, as could be expected, that Man is also *self-conscious*, and he is conscious of the spiritual centre of action in himself that characterises him as a person. Scheler denies animals self-consciousness, and here refers, rather than to biological evidence, to Leibniz. He explains that self-consciousness, the bending of the individual over his own existence made possible by his spirituality is the second hallmark of man. His spiritual force not only allows him to turn the environment into a world he understands and to know objectively resisting forces, but—and this is according to Scheler the most surprising of all—it allows him to understand his own physical and psychical constitution, and each of his psychical experiences and functions. That is the reason that man is the only being capable to make projects and to create his own existence.[11] This is nothing new: self-consciousness is the necessary and sufficient condition for human freedom. Like Husserl, he stresses that the mind of human beings is always *directed* towards *something*, it cannot but be directed towards a mental object. All phenomena are thus transformed into objects that can be grasped by the mind. One of these phenomena is for each man his "self." Human beings are capable of understanding themselves, because they do not, like animals, coincide with their feelings and emotions of one moment, but can grasp the continuity of their mental acts that constitute an object of thought.[12] It must be noted, however, that Spirit, as distinct from mental acts, is *not* an object. Spirit is the only form of being that cannot be an object; it is pure actuality through the free fulfilment of its acts. Its centre is the person, who is neither object nor thing, but a constantly self-accomplishing series of acts, of which the order is determined by essence. A person is *in* its acts and *by* its acts.[13] Only Man has a completed and perfected concept of things, of substances; animals never dispose of the object category the way men do, because they have no psychical centre, where all the sensations in different sensorial fields can be reported and by which they can be coordinated.[14] This difference is important, because this capacity is the basis for situating objects in space relatively to the psychical centre; men have an immediate concept of their own space, animals don't. The same reasoning is applied when time and space are considered. The conclusion is that only Man possesses from the beginning the concept of a totally empty space and empty time, or in other words of the absolute (ideal) form of them. Therefore, he situates the phenomena in these absolute dimensions of reality, in spite of the fact that science has taught us, as Scheler (wrongly) mentions, that absolute space and time do not exist. It is sad to see that, like so many other philosophers before him, he uses as evidence for his philosophical theses dubious examples of the animal world, but offers no serious scientific, in this case biological, arguments.

The philosopher investigates the essence of things, while the scientist investigates the phenomena as such. The separation of "essence" and "being there" is the most fundamental characteristic of Spirit and founds all other (unique) characteristics of Man. The search for essences requires a technique consisting of looking for the "logos" of things that have become objects of thought by peeling off their sensorial, concrete context. It is an attempt to substract from them their character of mundane reality. As we can see, Husserl's eidetic reduction is conserved, but interpreted by Scheler in a new manner. *In summa*:

> To be a man is to oppose a forceful "no" against this kind of reality. Buddha knew this when he said "it is wonderful to *contemplate* each thing, terrible to *be* it" [my italics], and when he developed a technique for the de-realisation of reality and of the self. Plato knew this when he linked the contemplation of the ideas to a turning away of the soul from the sensorial content of things and to a returning of the soul in itself, in order to find here the "origin" of things. And it is nothing else Husserl intends, when he relates the knowledge of ideas to a "*phenomenological reduction*," i.e. a "striking out" or a "putting between brackets" of the

random existence coefficients of the worldly things, in order to find their essences....[15]

Scheler analyses the negative aspects of the theories of Man stressed by Buddha, Schopenhauer and Freud, and concludes that though they all propose a negative conception of Spirit as the negation of the instincts and passions, Spirit is far more than that. Therefore, Scheler, instead of taking Spirit for granted on the contrary asks "what is it in man that enables him to sublimate the dark urge in him"? In the classical Western vision, which stems from the Greeks and from the Judeo-Christian religion, Spirit is Being; it *precedes* phenomenal reality and is at its origin, whereas Scheler is convinced that the Spirit becomes only real through the existence of human beings. This explains his world view: the stream of forces that can only produce "things that are merely there" and that are by chance like they are, this stream in the world in which we live does not flow downwards, but upwards.[16] As said before, the Spirit is a late, rare and fragile realisation, not the primitive ground of Being.

This is proof that though Scheler continues German idealism and though he is a Christian (and even for a number of years became a catholic, though later on he rejected Catholicism), he is in this respect, just like Hegel, at the opposite of the main Western tradition that goes back to Plotinus. This tradition gives meaning and sense to reality by picturing it as a circular process, starting with God, followed by the emanation of God, containing the World Soul, carrying the eternal ideas, the models of all individual things, with it. These are realised in the phenomenal world, and a layered reality is created of which Man is the summit. Man is the only being capable of reaching back for the ideal and the spiritual layer of reality, longing to be reunited with the origin of his being, God. Scheler turns this mystical vision upside down: instead of understanding himself as part of the Divine plan expressed in the reality that God created, the individual turns round, looks back and asks: "why is there a World at all?," "why is it that 'I' am at all?" The usual answer is that man transcends the rest of reality, because he is the reason of being of reality and because this reason is realised through him. For Scheler, on the contrary the profound relation of man to the ground and origin of the world is that this foundation understands and realises itself in him.[17]

Man has a task in this world, he must produce the Divinity, not, however, as men like Spinoza and Hegel thought, using only his reason, but first and foremost emotionally, by his active engagement as a person. Spirit and primitive urge must penetrate each other mutually, in order that God can come into being. This is the complete self-realisation of "that being that is through itself," namely God, and for this aim world history is a necessary process of gestation. By searching that being that is through itself and taking into the bargain the world as a "history," man becomes divine, and this is to be human, this is what lives in the human heart. Man is the only place of the becoming of God that is accessible for us and he is at the same time himself a genuine part of this transcendent process. All things are continually and at each moment produced from that being that is what it is by a constant creation through the functional combination of urge and spirit; only in man and his self are *both* attributes related to each other. Man is their meeting place and in him the *logos*, according to which the world is built, is at the same time an act that can be accomplished. Man becoming man and God becoming God are mutually dependent. Not any more than man can reach his destination without knowing himself to belong to both attributes of the upper layer of Being and without knowing that he himself is an inhabitant of this Being, not any more can that Being that is through itself reach its destination without the collaboration of man.[18]

In Scheler's world view, according to his beautiful expression, "Man weaves the tapestry of God." He realises that for many people this vision of a God that is incomplete, unfinished, becoming, is unbearable, but he replies that metaphysics is no insurance company for the weak, for people in need, but presupposes strong, courageous persons.

Teilhard de Chardin: an attempt to combine evolutionism and theology

At the end of the nineteenth century a confrontation between Christian belief and science was inevitable. In a more remote past, religion had the ascendancy over science and therefore had in all matters the last word. This situation slowly changed in Modern Times, when science by its rapid progress and its many successes gained such a prestige that it became more and more independent from religion. Its most important representatives became a sort of National Institution, and those who dared contest their theories were dismissed as "obscurantists." Indeed, it became very difficult to ignore scientific theories, let alone to combat them if they were in contradiction with religion. In the nineteenth century for the first time scientists felt no longer obliged to adapt their theories in order to bring them in accordance with the Bible and religious dogmas; on the contrary, their methodology openly forbade extra-scientific, metaphysical and religious presuppositions. As Laplace already put it in the presence of Napoleon, God had become a superfluous hypothesis, a statement to which the Emperor did not readily find an answer. Laplace could afford such a public pronouncement because he was a famous physicist, who had completed Newtonian mechanics, who was a successful mathematician and who had made progress in differential calculus and statistics. In addition to this, he was a member of the *Académie des Sciences* and for a short while minister of interior affairs, with the title of *Comte de l'Empire*.

In the late nineteenth century the Roman Catholic Church became aware of the threat of a dual society, a threat that was real, particularly in France, where after the French Revolution, the laicization of education had been a success. In the public schools, frequented by the majority of children of school-age, science was an important item on the program, whereas the clerics teaching in the catholic schools in most cases had no scientific formation whatsoever. The situation became very difficult for many priests, who could give no adequate answers to the numerous questions of their flock. The Church had to cope with the new situation, and Leo XIII (1890–1903), a pope who was more liberal than his predecessors, made an appeal to the clerical authorities to turn the tide by the creation of a "catholic science." The brother of this pope belonged to the influential Thomist movement, and the Holy Father put his hope in Thomas Aquinas, whose fame was to have successfully reconciled reason and belief. He accomplished this feat in the turbulent time of the Latin Averroists, who in the thirteenth century had gained great influence at the universities in the rebellious faculties of the artes liberales. What Leo XIII did not see, was that it was no longer the rationalistic philosophy of the Enlightenment that challenged religious faith, but nineteenth century science, an even more formidable enemy.

No catholic science was born, in spite of well-intended attempts to follow the directions of the Pope, by, i.a., the organisation of vast congresses that remained without result. With the next pope, Pius X (1893–1914), the Modernist Movement in the church came to an end and what has sometimes been called a Stalinistic period began, in which the Modernists were persecuted and a system of surveillance, snitching and arbitrary punishments was installed.[19]

It is against this background that the philosophical and theological anthropology of Pierre Teilhard de Chardin must be seen. He was a Jesuit priest who had studied geology and palaeontology in Paris. He spent fifteen years for his research in China, where he contributed with Davidson Black to the discovery of the Peking man. He participated in many expeditions all over the world and was respected as a specialist. However, he was concerned about the relation between science and religion, and this led him to seek a solution for this problem. His philosophical work is an attempt to realise a synthesis of a scientific and a Christian world view. The reaction of his superiors, however, was devastating: the diffusion of his work had to be prevented, and he was sent in exile to America. He was urgently exhorted to restrict his activities to research in the fields of geology and palaeontology.

Let us concentrate upon his cosmology and anthropology. Teilhard de Chardin, like so many Christian scientists, stresses the need of human beings to understand themselves as parts of a totality, and wants to start from what is scientifically established in the domains he is interested in, in order to draw from this the most general conclusions possible, and then use them for the construction of a global world view. In other words he wants to use science as a ladder to climb upon and to look over the walls it leans against. Of course this is an equivocal project, because the authority of the scientist can give the impression that his philosophical theories are scientifically valid. Specialists, from their side, generally do not appreciate the philosophical speculations of their colleagues if the distinction between their personal conceptions and their purely scientific statements is not made absolutely clear.[20] It is not this methodological rigor that Teilhard is seeking; on the contrary, he wants to make sense, to find the actual meaning of Christian faith, not *in abstracto*, but for men living in the first half of the twentieth century. He wants to take into account for this task what is scientifically recognised about nature and about the human species, and vice versa, he wants to make sense of scientific theories in the light of what he believes as a Christian.

As we shall see, his world view corresponds to the broad lines of Christian tradition, but he adds two characteristics to it: *cohesion* and *evolution*. God created only once, but in his initial creation all the possibilities are already present and wait for maturation. He starts his cosmology with the inorganic: in the beginning was matter, elementary particles adhering to each other, because from the first moment they contain the all-pervading principle that he does not hesitate to call "*love*." All that exists seeks cohesion and more and more complex wholes are formed that together form still more complex totalities. Descartes too believed that matter and movement are created by God, that matter obeys the laws of nature and has the qualities required to give rise to all the elements of reality and their relations, but his vision is rational and cool. Teilhard on the contrary is a poet and writes an ode to matter that has become famous:

> Be blessed, dangerous Matter, violent sea, untameable passion. Be blessed, mighty Matter, irresistible Evolution, continually born Reality. Be blessed, universal Matter, Duration without limits, boundless Ether, you that, overcoming and dissolving our narrow measures, reveals to us the dimensions of God. I greet you, divine Midst, laden with Creative Power, Ocean agitated by the Spirit, Clay kneaded and animated by the incarnated Verb.[21]

Each particle of matter, each atom, each molecule, each lump of matter has an "exterior" and an "interior" aspect, the latter conferring to it a certain degree of consciousness, proportional to its complexity. It is this primitive consciousness inherent in all things that is the original creative force that drives evolution. It explains the origin of life, the transition between the pre-biotic and the biotic phase, though the details of the historical process of the beginning of life will never be known. Therefore there is no "point zero" between non-living and living matter. Teilhard warns us, right from the start, that each transition is at once a step in the evolutionary process and therefore is a continuous development over a span of time, but also and at the same time the discharge of accumulated inner energy, causing the emergence of something that did not exist before and is *qualitatively* (read metaphysically) different. This principle, which is important for the theological aspect of his cosmology, is what he terms "discontinuous continuity": like in a metamorphosis, the concentrated psychical interior force enables matter to take an exterior form that matches it. Once life had come into being, biological evolution could start, the species could develop. If a new form can only grow if the energy is concentrated enough, this means that the interior force always has precedence over mutations. Thus there is built in a guarantee that evolution is *neither merely a mechanical process nor a historical process* based upon the interplay of chance elements, but that it is the expression of the potentialities God bestowed matter with and therefore that it is ultimately the expression of the Spirit or Divine Providence. The biosphere is an organic progressing totality, sensing the possibilities of development into new forms by trial and error processes and taking each time the good direction, i.e., the course that is in accordance with the divine plan. Thus he combines evolutionism with what he must have considered to be an orthodox religious view.

In *Le Phénomène Humain* (The Human Phenomenon), Teilhard uses his palaeontological knowledge to sketch the evolution of the tree of life. Though he accepted evolutionism as the only scientific explanation of the origin of the species, which like all natural phenomena had to be conceived as a spatio-temporal process, it was not yet clear to him that neo-Darwinism was generally accepted and Lamarckism rejected.[22] Teilhard was not a biologist and did not pretend to be one; in fact as Jean Rostand has observed, he considered the discussion of the true nature of the evolution of species as a subsidiary question, given that the process was from the beginning *directed*. His transformism is superficial and confused; he neglects the biology and chemistry of the cell, he neglects genetics, he is not interested in chromosomes or genes. In the theory of evolution of Teilhard de Chardin, writes Rostand, we can find a bit of everything, natural selection, mutations, i.e. neo-Darwinism, as well as a curious psychical Lamarckism. The latter is reflected in his idea that the form of an animal is derived from his instincts: the carnivores under the mammals have developed their strong canines and claws, because they had the soul of a carnivore, the runners under the mammals have long legs, because they had the soul of a runner. The same can be said of the swimmers, the gliders, the burrowers.[23] Moreover, following Teilhard, acquired characteristics, above all psychical characteristics, are transmitted by heredity:

> Qualities and deficiencies in the individual develop with age. Why,—or rather how—wouldn't they also not be accentuated phylogenetically? And why, in these dimensions, wouldn't they react upon the organism to knead it after their image?...[24]

The last and most important "leap" in the evolution of species is the advent of Man. Again, though the process is evolutionary, there is at the same time a qualitative "jump," as important as that between the pre-biotic and the biotic stage: consciousness has become self-consciousness. This is the moment the "noösphere" (sphere of the mind) has come into being, beyond and above the biosphere, namely the sum of all human consciousness. Man is not the result of a special divine intervention, but the outcome of the evolutionary process God initiated at the beginning of time, in him a new layer of reality is born, the crown upon creation. A new era begins; the Earth has found its soul.

We must remark that here again "discontinuous continuity" is supposed, which enables Teilhard to circumvent the thorny question of the human origin(s). As a palaeontologist, he cannot ignore it, but he must tackle the problem whether originally different species of men coexisted, or whether there was only one species from which all men have evolved. And above all, he has to find a solution for the biblical statement that a first couple of men must have existed, that we all descend from Adam and Eve and inherited from them Original Sin, without which the sacrifice of God the Son and our redemption by him has no meaning. Teilhard has taken his precautions: from the start he has warned us that evolution is not a multi-stage process, but gradual, and that we never can determine exactly which individuals belong already to a new species and which ones still belong to the original species. This leaves room for interpretation. In *Le Phénomène Humain* the footnotes are of great importance, containing often a self-justification in anticipation of criticism. In one of those concerning the problem of the birth of man he writes:

> Which comes to saying that if the science of Man cannot affirm anything for or against *monogenism* (only one initial couple,...) it on the contrary pronounces itself firmly, it seems, in favour of *monophylism* (only one phylum).[25]

This universe described by Teilhard de Chardin is a spiral with a broad basis, rolled upon itself and slowly unrolling itself, while producing ever new forms of being in such a manner that at the same time all these forms of being grow in complexity and in mental capacity and that the spiral converges by these two aspects in Man, and beyond Man into a "*Point Omega*." This ultimate kind of being, God, is no longer God the Father and the Spirit, but has a human face; it is the incarnation of God the Father through whom the Spirit works, Jesus Christ. Teilhard dreams of a unification of mankind, of Man at the dimensions of the Planet Earth. His prophecy is that after matter has met itself in Man, the consciousness of individuals will be surpassed in a super-personal consciousness, and will concentrate in the cosmic focus, Point Omega or the End of the cosmic evolution, wherein all the initial forces are pulled together and converge. The love of Jesus Christ is already present today, purifies mankind; inspires mankind to reach the final goal of evolution, God.

Instead of trying to prove that man is an exceptional being and no longer an animal, by collecting anthropina (unique human characteristics), such as the alleged fact he has poor instincts, his neoteny, his plasticity, his historicity, his free will, his symbolic language, his ability for abstract reasoning, etc., Teilhard founds the fact he is unique in his relation to the Universe and his Creator. Man is, of course, unique by his spirituality that manifests itself in his achievements, in his civilisation, collective memory, art, etc., but above all man is special because he is the aim of the cosmic evolution, because he has been planned from the beginning; he is unique because of his role in the universe and because of the task he has in completing the evolution of the world. This is why self-consciousness that is nothing else but the quintessence of the force driving evolution is essential. It is only in being aware of his own humanity and the direction of the vast process he is part of that Man will be able to understand the true nature of reality and his task in this world. He can only accomplish his assignation if another condition is fulfilled: the individual must surpass himself, transcend his individuality in order to participate in the planetary mission of Mankind and in order to contribute to its returning to its origin, to the Beginning and the End, Alpha and Omega, God.

This conception of reality is in keeping with the millenary Western world view inspired by the Christian version of Plotinus: God has created the world, poured himself out in it; the divine Spirit is present in it and moulds the multitude of forms reality can take, which are hierarchically ordered into different spheres. Man is the ultimate sphere, the spiritual being that can establish a relation with God and can complete Being by his works, close the circle and realise the Kingdom of God on Earth. The variant of Teilhard is that the different layers are no longer the realisations of ideas contained in the World Soul, the "*nous*," the Holy Spirit, but that they evolve from the primeval matter God created in the beginning. Thus, matter containing the germ of spirituality right from the start, Man is not by his spirituality the anti-nature, but the highest form its evolution can take. Consciousness and self-consciousness are crucial in this world view, but they are not contrary to materiality, but inherent in it. This is one of the features of his philosophy that make it truly religious; the world is one, it is not alien to us, but we are a genuine part of it, we are united to it by the universal principle of Love, we have our reason of being in it.

Max Scheler's vision is that of a philosopher who is religious. For him it is with Man that the divinity comes into the world: "Man weaves the tapestry of God." Teilhard is a Catholic priest, who is also a scientist; to him God must be, not only omega, but also alpha, *before* the phenomenal reality, as his Spirit pervades his creation. His world vision is an attempt to bring as much as possible his faith and his knowledge into accordance. That Christian religion contains the Truth is unquestionable for him. As an anthropologist he has been in contact with many different cultures, but he never doubted that Western culture is superior to all the others. He does not say it explicitly, but it is clear that to him the reason for this superiority and its proof is that it is the cradle of Christianity. It is in the culture of Europe that all human phenomena discovered in other parts of the world really find their final development.[26] Man, and in the lead European Man, has a cosmic task, a moral duty: the completion of the evolution of reality. For the individual, however little he can contribute to this goal, this vocation gives him a resource for transcending the dimension of his short existence and to make sense of it.

But what about the fate of humanity, will it not one day in the future be destroyed by a great disaster, or by a cosmic event, a gigantic collision of our planet with another celestial body, or the exhaustion and death of our Sun? Man is part of the divine Providential Plan; far from Chardin to consider, like Nietzsche, the possibility of the extinction of the human species, or the possibility that perhaps it will give rise to a completely different species. Jean Rostand cites the following text from *Le Phénomène Humain*:

> I believe to be able to say—, by basing myself upon what all the past of evolution learns us, that we do not have to fear one of the multiple disasters. Though they are possible in theory, we can be certain, for a superior reason, that they will not take place.[27]

And this is the comment the atheist Rostand gives:

> Never, in reality, one has gone so far in the anthropocentric and "providentialist" optimism. Thus, following Teilhard, Man cannot finish before his hour has come, not even stop or deteriorate, and this simply because he is Man, an irreplaceable creature, and because there exists a secret complicity between the Immense and the Extremely Small on the one hand, and human conscience on the other. Man is the ally of the Universe, which cannot abort, and we can count on this protection, this tutorial vigilance. Does this Teilhardian "super-optimism" not express a need of security that in fact gives away a profound anguish? (...)
>
> For Teilhard, the human condition would be too desperate, too unacceptable, if it would reduce itself to what the unbelievers see in it, namely to its first appearance. And on this point, I do not feel myself very far from Teilhard... Like him, and so many others, I feel the distress of the human individual in face of death that will annihilate all of his effort; I feel the smallness, the misery of this adventure bound to become defeat and nothingness; I feel the insufficiency of this world, where nothing survives, where nothing lasts, where nothing counts...
>
> The sane, virile attitude of wisdom and reason is not to judge the fate reserved for the human being to be horrible or absurd. Our sensibilities are wrongly made, ill-adapted; they are mistaken if they revolt against what is. One should do better than to be resigned, even if it is with stoicism, to the human condition: one should come to find, by intelligence and heart that in order to live and to be a man there is no need for stoicism or resignation.[28]

Existentialism: Jean-Paul Sartre's extreme subjectivist anthropocentrism

The philosophical anthropology of our time can be considered as an answer to the problems posed by the progress of science, especially in biology. Sartre proposes an interesting view on man, but in contrast with the German philosophical anthropologists and with Teilhard de Chardin, totally ignores the science of his day. When treating of human individuals he ignores the implications of neo-Darwinism, of the biology of the cell, genetics and animal psychology, all of them well-established. In keeping with the idealist tradition, which goes back to Socrates, he is not interested in nature, but only in one thing: man and his consciousness and self-consciousness as the sources of the most precious human characteristic, freedom. There are of course several reasons for this, amongst which his peculiar personality and the intellectual confrontation he had in his early years with German philosophy.

Biographical data are necessary in order to situate the work of philosophers in the context of their own time, and in most cases they can be useful for a full understanding. Where Sartre is concerned, we can say that his work itself is autobiographical. Existentialists in general are convinced that philosophy cannot be an academic matter, but must be based upon existential experiences. Though Sartre is not an exception, nobody went as far as he did: his life is an exemplification of his philosophy, or rather the reverse, his philosophy is an abstract generalisation of his existence. In other words, it is not wise to try to understand his theories like one would try to understand those of Descartes, Hume or Kant, because they are neither based upon usual rationality nor upon common sense, nor upon empirical principles, but upon his personal experience.

The Anglo-Saxon mind is baffled by this kind of thought; starting from the wrong presuppositions, Sartre's work seems incomprehensible. Taking it for something it is not and judging it using standards that cannot in principle be applied to it, he is dismissed as not being a "serious" philosopher. For the sake of simplicity he is reduced to a kind of journalist with literary talent, an appreciation that, ironically, would not have displeased Sartre. John Passmore in his classical *A Hundred Years of Philosophy*, writing a survey of contemporary philosophy, and conscious of the great influence of Sartre's thought on the continent, he makes an honest effort to analyse and explain his work to his readers, but must conclude:

> However, not only English readers of Sartre feel uneasy about his work. Paul Foulquié, wrote a volume on existentialism in the famous series *Que sais-je?* of the Presses Universitaires de France, in which he devotes much of his attention to the work of Sartre, assessing it critically.[29]

Like Passmore, Foulquié goes along with Sartre and tries to establish a kind of empathy with him, but at the same time checks the rationality of his arguments. It becomes clear that many times, forestalling possible objections, Sartre makes use of paradoxes instead of giving a rational answer to the problems that occur. Foulquié cannot but conclude that his conception of reality is absurd:

> This may be as it is, but is not the putting between brackets of the rationality of reality much more at the opposite of the philosophical attitude than the putting between brackets of one's existence? A philosophy of the absurd cannot claim to be a philosophy.[30]

Why is this type of philosophy interesting all the same? In the first place, because of its immense influence on the culture of Western Europe. We must not forget that, more an attitude to life than a rational theory, existentialism shows a great diversity; it appeals to Christians as well as to atheists. Moreover, in the existential situations and experiences it refers to, most people can recognise aspects of their own existence, which gives them an impression of the veracity of this philosophy. Whether we read Heidegger, or Sartre, however different they are, psychologically many of us can immerse themselves in what they have to say to us, if only we forget for a moment critical, rational, logical argumentation. They offer us no ethical system, not even a moral philosophy, and yet they have a message to deliver that can direct our lives. In the second place it is by forming an idea of what they actually say to us that we discover the sources of the picture of man that the majority of the intellectuals of Western Europe cherish. It is not difficult, in spite of its air of novelty, to discover that whether under its Christian form or under its atheist form, existentialism is in broad lines the continuation of the old metaphysics of the West.

Why did we choose Sartre to exemplify existentialism? The reason is that in his oeuvre two major themes of Western tradition are central, namely the human being as a *subject*, i.e. as consciousness and self-consciousness, and as *liberty*. In fact, he makes a blow up of these concepts and presents them as constituting the humanity of man and as the precondition of true being. In his ontology he answers the question "what are consciousness and liberty?" and in his existential psychoanalysis he investigates how we experience them.

Sartre has been an outstanding and even an unsurpassed biographer, who wrote also exceptional autobiographical works. His novel, *La Nausée (*Nausea*)*, recounting in a literary transposition his personal experiences, is by most commentators considered an excellent starting point for understanding what his philosophy is really about. However, I prefer *Les Mots (*The Words*)*, because Sartre who was more than fifty years old when he wrote it, does not express his philosophy in it, but looking back at his intellectual attitude puts it into the perspective of his psychological development during his childhood and adolescence. Indeed, if, like Sartre would have wanted it, we must understand his philosophy from what he has been as a person, it is here that we can find the profound reasons for it.

The crucial experiences of Sartre's childhood

A year after the birth of Jean-Paul, his father, a naval officer, died of intestinal fever. His mother, Anne-Marie Schweitzer, returned to her parents, and lived with them in Meudon and in Paris. Though her parents accepted the situation, she felt, because widows received only slightly more consideration than unmarried mothers, as if she was in a way to blame for the sudden abortion of her married life and had to make it up to her parents. She kept their house, being at once a governess, a nurse, a lady-companion and a servant. Her father, Charles Schweitzer, of Alsatian origin, was a respected teacher of German at the Lyceum. He had reached the age to be pensioned, but in order to support mother and child he continued to teach. A man of the nineteenth century, he always had shown a propensity for the sublime. Now that he grew old, though he had never before shown a great affection for his children, he became extremely fond of his grandchild. In the analysis of Sartre he became a kind of paradigm of grand-fatherhood. His wife, Louise Guillemin, a witty but cold woman, was not happy in her marriage. She had become over the years suspicious of others, sceptical, and an unbeliever. Having nothing to do, she read novels most of the time. Though apparently the Schweitzers formed a model family, they all played a role: the solicitous grandparents or the twofoldness "Karlémami" (Charles and Mammy), Anne-Marie, his overprotective mother, who was more of a sister to him, and Jean-Paul himself in the first place.

He did not respect the grown-ups, he merely played that he had become what they wanted him to be, a model child, always good, intelligent, precocious, etc. Having no father to give him an identity, he became what the others thought he had to be. Extremely intelligent, he knew that he was play-acting, but at the same time, he half believed to be the character he played. In reality, having no point of reference of his own, he was nobody. There was no real parental authority; he was free in an almost absolute sense. Nevertheless, the only thing he had in fact chosen was to conform to the idea others had of him and thereby he was most unfree.

His mother let him keep for a long time his long curly hair, and, only when Charles protested that he did not want a girl for grandson, she reluctantly had it cut. He naturally believed he was cute, but when growing up he discovered he was ugly: he was, like his father and the other Sartres, short; a white spot grew on his cornea, making him blind in one eye and squinting. Growing up between adults, he was extremely lonely during his childhood: "till I was ten years I stayed alone between an old man and two women."[31] When he went for a walk with his mother in the park the children did not want to play with him, he was not capable of making social contact. No wonder, that the child when five years old had a serious nervous depression and was haunted by death.

In fact, his situation was not that peculiar, he lived the "normal" life of a child in a bourgeois family. The family was not only insincere in their mutual relations, but as it was common in bourgeois milieus, in its ways of thinking as well. They were conformists and went occasionally to church, but nobody was religious in any genuine sense. His grandfather, says Sartre, was too much of a comedian, not to need a Great Spectator for his performance, but he never thought about God, except on the crucial moments of life. He was certain that He would be there on the moment of his death, but, being brought up in the Protestant tradition, he ridiculed Catholicism whenever it pleased him. To Sartre's mother faith was a consolation, but a strictly private affair; she kept her religious feelings to herself. Perhaps the belief in God could have given the boy an identity; God could have given him the necessity he lacked by having wanted him, by having chosen him. Perhaps he could have found his vocation in saintliness, but his grandfather's ironical remarks prevented him to give in to this temptation.

The grandmother read a lot. The grandfather was a learned man, possessing a rich library. Jean-Paul being fascinated by these strange objects called books the adults liked so much, had taught himself to read, a feat the family was extremely proud of. When the little genius went to school, he was not put in the first, but in a higher class. However, to the general dismay it was discovered that instead of being in advance of his age, the child was behind, having learned to read, but not to write correctly. His orthography was awful and his grandfather obtained from the director of the school that during playtime he could stay in the class under special surveillance to catch up, which isolated him further. It is only when he went to the Lycée Henri IV that he had normal social relations with the boys of his age and made friends. For the first time in his life he was no longer superfluous; he was treated as an equal and he felt wanted, necessary as a member of a group.

It is remarkable that Sartre in this story of his childhood does not mention either ordinary objects, or bodies (except books). The description of his mother is psychological, we do not know what she looked like, but only that she was tall. His grandfather was very impressive; he could pass for Jehovah when his hair was still black, for God the Father when his hair had grown white. He dressed elegantly and according to the fashion of the time; he was photogenic, and that is all he mentions. Things are nothing to him, the real world consists of psychological relations and, from the moment he can read, of words. He devours anything he can lay hands on, without any order or system, for example the *Larousse Encyclopédique*. In fact, he knows what exists, not from his own experience, but from the articles and illustrations in the Larousse. What is real is not this thing or that thing, but its definition, its essence. He acknowledges that he was a Platonist; ideas were given to him as things, and he has remained in this state for a long while; it took him in fact thirty years to shake it off.

> A Platonist by state, I went from the knowledge to its object; I found in the idea more reality than in the thing, because the idea gave itself as a thing. It is in the books that I have met the universe: assimilated, classified, labelled, thought, but still frightful; and I have confounded the disorder of my experiences from books with the hazardous course of real events. Here the idealism comes from which it has taken me thirty years to get rid of.[32]

His imagination is vivid and he starts writing, realising in his tales the things he dreams of and stuffing them with learned details he transcribes from the Larousse.

The boy starts a quest for a mission that would save him from his nothingness, his utter contingency. Invariably in his tales he is the hero who enters on the scene at the crucial moment when he is needed and who makes all well. This written daydreaming replaces real life. When this literary activity is discovered by the grown-ups, he has found a new vocation: he will become a writer. Though his path as a writer was in the beginning not strewn with roses, nevertheless this is what saved him.

Sartre's studies

Sartre studied at the *Ecole Normale Supérieure*, and in this period philosophy was central in his field of attention. He had read Henri Bergson when still at the Lycée and he became interested in the work of Husserl, when he was a teacher in Le Havre (from 1931 on). It is in this period that he wrote *La Nausée* (Nausea) his first novel to be published in 1938, by Gallimard. Till 1939, Husserl's philosophy remained all-important to him. He requested at the instance of his friend Raymond Aron a sabbatical leave, and in 1933 he spent with a study-grant a year at the *Institut Français* in Berlin, where he thoroughly studied Husserl's phenomenology. Sartre could read German, but was not fluent in it. Apparently, in contrast with his Jewish friends at the Institute, he was not interested in the political situation in Germany. He did but one thing: work. He stayed mostly in the restricted circle of his compatriots and fellow students and seemed unaware of the danger Hitler's election represented and the serious threat to freedom Nazism represented.

Sartre's ontology

Securing its foundation

It is only in 1938 that he really discovered Heidegger, studied his work in detail, along with the work of Hegel and of Husserl. These are the "three H's" that are the main source of Sartre's inspiration; he transforms their thought into a philosophy of his own, rejecting both idealism and materialism.

In the first place he denies that Husserl's phenomenology, which will be his starting point, is a form of idealism, which he identifies clearly with Platonism. The essences revealed in Husserl's *Wesensschau* (speculative search for the nature of things) are not absolute; they do not exist in a world apart like in Plato's philosophy, but only to the extent that the phenomena present themselves as intentional objects to a subject. This subject must previously have been cleansed of all presuppositions that could hinder the intuition of the essence of its objects of thought. This is not psychology in any ordinary sense, but it is *reflexive* psychology: the subject grasps itself as a "pure self" and then by a speculative process discovers through the structures of this consciousness the nature of things.

In the second place, Sartre, like Heidegger, comes to believe that Husserl's eidetic reduction is in fact superfluous; both replace again the "pure ego" obtained by Husserl by a rich consciousness that is not stripped of its idiosyncrasies, of its emotions and relations to the objective world. Thereby, at least in their opinion, in their respective philosophies man has become concrete again. It may seem awkward to call a consciousness concrete, but as we shall see, Sartre did not conceive of it as a substance different from the body and from the behaviour it shows, not as an empty box that could be filled with any mental object whatsoever, but as coinciding with the totality of the individual.

In the third place, he is confronted with the thorny problem of solipsism. Husserl identified knowledge of something with being, and therefore he could not avoid solipsism: if what exists is what I know to exist, my knowledge being *my* knowledge, what exists is what exists *for me* and no consensus is guaranteed.[33] Hegel found a first remedy for the problem by declaring that if someone discovers the existence of another consciousness this other consciousness thereby becomes for him an object of thought. On the other hand, the other not only discovers himself as what he is as an object of thought for someone else, but vice versa in turn he approaches the other as an intentional object. Thus, each self is a being for himself only through being a being for someone else.[34] Nevertheless, being is here still the same thing as knowledge, though now it is anchored by the knowing subject in another consciousness. Heidegger finally thought to have found the solution: human beings are in the world, they are real beings and their relation is a relation of being, not of thought. This is the starting point for a new definition of what it is to be human.

We have a *pre-ontological intuition*, i.e. immediate certainty of our being ourselves. The self is by its very essence always directed towards what is not itself, and therefore *this "other" must exist, or I could not exist.*[35] Of course, this does not solve anything, because it still derives speculatively being and essence from what is given in our consciousness, but Heidegger does not recur to common sense and ordinary reasoning. He simply declares that intentional objects that are immediately given in our consciousness, such as Being, Nothingness and our Selves, are revealed in crucial existential experiences as essential to that consciousness. Sartre adopts a version of this point of view.

The subject

The foregoing is the basis for Sartre's philosophical construction of the subject. He declares that it is different from all the other elements of reality. We know what Being is, namely merely the condition for the apparition of the phenomenon. The phenomenon is by definition something that appears to somebody; but Berkeley's *esse est percipi*, to be is to be perceived, is not enough. The question is how do we know this somebody? In the empirical theory it is known as an apparition to itself. This is a vicious circle that must be broken by supposing, in keeping with Husserl's idea of the transcendental ego (the ego that reflects on itself), that the subject is absolute. It is given to itself, not like all the rest that appears to it, but in a special way: it is a kind of consciousness of consciousness that is an immediate, non-cognitive relation of the self with the self. Sartre claims there is a pre-reflexive ego that is the condition of the Cartesian cogito.[36]

Being

We must take notice of the fact that the Kantian noumenon has been discarded in phenomenology: reality *is* what appears to us. For Kant what appears to us is that what is transformed by us into a *representation,* but its reality is independent of human consciousness and cannot be known *as it is in se*. Sartre, in trying to avoid idealism, reverses this reasoning; reality *is* what shows itself to us; without us it is an indistinct indeterminate mass. It is by appearing to us that a mountain *becomes* a "mountain," a river *becomes* a "river," a fish *becomes* a "fish," etc. This conception is also different from that of Hegel, who believed the world, as it is, is already potentially ideal and rational in se, because it can be thought. Man brings to light the rationality of the world and thereby he is an instrument for the realisation of the *Weltgeist*, the spirit in its most general sense, pervading reality. Husserl, from his side, believed that we, humans, do not "make" the essences, like Aristotle supposed, but "find" them in our transcendental ego, our consciousness of our consciousness, once it is cleared from what makes it impure. This is the source of an "objective" idealism. From Sartre's point of view, the order in the world is not preordained, but on the contrary reality is completely absurd without the intervention of man, who does not *find* this order in the structure of his own mind, but *creates* order and meaning. Thus, to be is to be *experienced* by human individuals, who, like Leibniz's monads, mirror the world and are mirrored by each other: *man is the precondition of what is real.* Thereby Sartre turns upside down traditional Western ontology: not God, not Being, is the founding principle of all that exists, but human subjectivity is the source of reality. Thereby reality itself, true being, is no longer absolute, but it depends upon a more primitive reality, the ego.

Nothingness

As we have seen, Sartre has established what Being taken in its most general sense is: Being *is*, Being remains *in itself*, Being is merely *whatever it is* when it does not appear to a subject. In fact it is a mere precondition for the phenomena.[37] Now he must still conjure Nothingness from his hat. Again it is in the subject that we can find the solution. Like Heidegger, Sartre rejects the logic of common sense, which declares with Parmenides "being is and nothingness is not." He posits that in order to be able to know and claim that something is not, non-being must exist, or, in other words, Nothingness must pre-exist before we negate. Again, miraculously, it is given to our consciousness, as a peculiar immediate experience. Whereas all other living beings and all the elements of nature are simply what they are, not being there *for* themselves, we are always to ourselves *what we are and at the same time are not.* The principle of identity does no longer hold where human reality is concerned, and therefore we are a source of Nothingness. Like a true scholastic, he formulates it thus:

> To grasp the problem from close by, it must be recognised that we cannot concede that Nothingness has the property "to annihilate itself" ("se néantiser"). For, though the verb to "annihilate itself" has been conceived to take from Nothingness even the last seeming of being, it must be admitted that only *Being* can annihilate itself, because in whatever way it is done, in order that self-annihilation takes place there must be Being. But Nothingness *is not.* If we can speak about it, it is because it has only an appearance of being, a borrowed being, as we noted previously. Nothingness is not, Nothingness does not annihilate itself, "*it is annihilated.*" This leaves us with the fact that there must be a [kind of] Being—that cannot be the being coinciding with itself (*l'En-soi*)—and that has the property to annihilate Nothingness, to support it from its [own] being, to sustain it continually by its existence itself, *a being through which nothingness comes into things.* But how this Being must be related to Nothingness, in order that by it Nothingness comes into the things?[38]

Sartre then proceeds to demonstrate that it is unthinkable that this being, namely man, would either actively or passively create a transcendent Nothingness extraneous to himself, and therefore: "... *the being by which Nothingness comes into the world must be its proper Nothingness*."[39] In nature there is not the least gap between cause and consequence, but in human consciousness there is Nothingness between the present and the immediate past. On the very moment that I am conscious that I am this or that, I am already something else. The reasoning is dialectical and can be found in Hegel's *Wissenschaft der Logik* (Science of Logic). Being cannot be specified; it is the emptiest concept, containing in itself its contrary, namely Nothingness. Both moments, the thesis and the antithesis, produce their synthesis, Becoming, at the same time remaining present in it, though on another level. This faculty human beings have to distance themselves from themselves, containing being and non-being while they are becoming, gives them the chance to choose what they want to become and this is the freedom that distinguishes them from the beasts. In Sartre's words:

> Therefore the conscious being must constitute itself in relation to its past as separated from this past by nothingness; it must be conscious of this gap of being, but not as a phenomenon that it undergoes, but as a structure of the consciousness that it is. Freedom is the human being taking its past out of the game and secreting its own nothingness.[40]

Sartre has not only established that there is Being, the precondition of existence, and that there is Nothingness, but he has also—like Baron von Münchhausen, who once fell into a bog and was in danger of drowning, but pulled himself and his horse out of it by the tail of his wig—secured his own subjectivity as the source of a unique property: the free creation of reality. In this extraordinary view it is not Being that lends man his essence and meaning, but the other way round.

Nature

Thus, Being is not at all the forever receding horizon we long for, the "Urgrund" or primitive ground, as it is for Heidegger, but it is rather that with which humans have no common measure, that which is utterly alien to them, that which coincides with itself, exists *being there*, but does not exist *for* itself. There is no reason at all why there is Being rather than Nothing. Sartre says in *Les Mots* that, like Socrates, he is not at all interested in nature, but only in man. However, there is more: he is *disgusted* by nature, by Being, insofar as it is unrelated to man. This reminds us of Hegel's aversion of pure materiality, which is alogical and therefore evil. In this respect the famous passage from *La*

Nausée, where Roquentin, the central figure, on one peculiar moment experiences nature in its quality of pure Being, is revealing:

> So, I was a little while ago in the public garden. The root of the chestnut tree bored itself into the soil, just under my bench. I didn't remember it was a root. The words had vanished, and with them, the meaning of the things, the instructions for their use, the feeble clues that men have traced on their surface. I was sitting, a little bowed, my head low, alone before this black, knotted mass that was totally brute, and that frightened me.[41]

Being has lost the varnish of the conceptual order men impose on it and has become the contrary of man, Nature:

> The chestnut tree pressed itself against my eyes. Green rust covered it halfway its height; the bark, black and swollen, resembled boiled leather. The little sound of the water of the Masqueret fountain flowed in my ears and made its nest in them, filling them with sighs; my nostrils ran over with a green and putrid smell.[42]

What his senses tell him is that nature is absurd, it makes absolutely no sense and by its sensuality defiles him. In fact, nature is not *justified*; it is there, it is factual, but it is contingent, it could perfectly well have been otherwise or not at all. As we have seen, there is no answer to the question why Being is rather than Nothing, and no reason why *this* Being is, rather than a totally different one. Roquentin is exasperated by the utter redundancy of mere existence, and he loathes those tranquil people minding their daily business, accepting things as they are, taking them for granted.

Science seems to explain satisfactorily the state of affairs in reality, by showing *how* things come about, by discovering regularities, formulating laws. However, all this is futile; it is beside the question, because it does not tell us *why* things are as they are. Roquentin imagines all these content, self-satisfied people the day an alternative reality would take over:

> In the mean time, the great vague nature has slid itself into their town, it has infiltrated itself everywhere, in their houses, in their offices, in themselves, they are in the midst of it, they breathe it and they do not see it, they imagine it is outside, twenty miles from town. I *see* it, this nature, I *see* it… I know that its submission is laziness, I know it has no laws, which they take for its constancy… It has only habits and it can change them tomorrow.[43]

Nature is contingent, not because God created it, and not only could have made it otherwise and at all times can intervene in it, it is not because it is the expression of the absolute power and the free will of God, but it is as it is *for no reason at all.*

Crucial in Sartre's philosophy is this contingency and degradation of Nature; it is at the opposite of Heidegger's idea of transcendent Being beckoning us at the horizon and containing the Truth about ourselves. Empedocles described the chaotic state in Nature in the beginning, when the earth and the sea teemed with extravagant forms of life created by chance. Some of these monsters that came into being were not viable, others survived and eventually order was established. In Sartre's vision the reverse takes place: science tells us that Nature operates in obeisance to laws, and though at first sight this may seem convincing, for those who are not content with superficial appearances it is obvious that, on the contrary, it is undetermined, unpredictable. And if, indeed, all of a sudden Nature changed it habits, became unreliable, chaotic, full of monsters? The rich though morbid imagination of Roquentin runs away with him. Nature is *in ovo* a nightmare: a family man is walking in the street and sees a red piece of cloth, blown away by the wind. It approaches, and the man can now see what it really is: a chunk of putrid meat, soiled with dust, maggots crawling all over it, dragging itself along, rolling in the gutter, ejecting spasmodically gusts of blood. Or in another example he imagines a mother discovering a sore on the cheek of her child and asking "what is it that you have there?," and examining it sees it swell, split open, and then a third eye appears. Another man, feeling a tickling in his mouth, looks in the mirror and opening his mouth sees that in the place of his tongue has grown a millipede, wriggling its innumerable feet and scraping his palate. He tries to spit it out, but it is part of him and he will have to tear it out with his own hands.

> Then I shall burst in a laugh, even if my body is covered with dirty, suspect crusts that open themselves and become flowers of flesh, violets, buttercups. I will lean against the wall and I will shout at them as they go by: "What have you done with your science? What have you done with your humanism? Where is your dignity of *roseau pensant* ('thinking reed'?)."[44]

In a world that is not made following a perfect rational plan, in spite of its seeming lawfulness, there is no necessity, and man being part of this world, is not necessary either. Neither his presence in the world, nor his rationality, nor his moral values are guaranteed by a transcendent principle.

The idea that no intrinsic sense is present in the world is not strange for an atheist, but the disgust that Being in general and especially Nature inspires Sartre is unique. It goes together with a great deal of anguish, with perpetual questioning of the most normal things and with the idea that if the world is absurd, to be part of it is to be absurd too. The only way to make up for our presence is to *merit* to be there one way or another. Sartre as a child was *a-normal*, literally outside the norm; in a way he was not a child like the others. Probably his very special psychology, his very special approach to reality, his personal experience, was the source of his belief that somebody is only authentic, if he is conscious of the profound lack of sense in reality and of the necessity *to justify* himself by transcending his "merely being present."

If Freud derived his general theory of the human mind from the study of the neurotic minds of his patients, Sartre rather constructs his theory of reality upon the idiosyncrasy of his own encounter with reality. Little Jean-Paul did not feel like most children that it was good to live, on the contrary, he was depressed, morose, he lacked *joie de vivre*, the spontaneous joy to live. Roquentin, who resembles in many respects Sartre himself, is not only disgusted by the tree in the park and by all those ordinary things other people unreflectively take pleasure in. They not only sicken him, but he is even revolted by his congeners, who in their primitive innocence are content just "being there":

> All things, softly, tenderly, let themselves go to existing, like those tired women who give themselves over to laughter and say: "It is good to laugh," with a wet voice; they spread themselves out in front of each other, they made abject confidences about their existence. I understood that there was not something in between inexistence and this enraptured abandonment. If one existed, one had to *exist till there*, till the mould, the distension, the obscenity.... Redundant: it was the only relation I could establish between these trees, these fences, these pebbles.[45]

What is truly human in us is that we are *en soi* and *pour soi* and struggle to become a perfect *en-soi-pour-soi*; truly human is to choose ourselves and thereby gain our own necessity, to become godlike or *causa sui*. Though incessantly we are striving for this, and we cannot but pursue this goal, we never attain it. In fact, our consciousness is an ailment, a curse; it is the strange and paradoxical capacity to be what one is and at the same time not to be what one is. Nevertheless, it is here, in this wondrous ability of man to contain being and nothingness at the same time, that Sartre seeks the remedy for his disgust with reality.

He tries to count the chestnut trees, to situate them in respect to the kiosk, to compare their height with that of the maple trees, but all in vain: they still merely exist unreflectively, there is no necessity in their existence that thereby is nearly inexistence.[46]

The body

Before we shall be able to come to a conclusion about Sartre's conception of man and his humanism, we have yet to consider how he conceives human materiality, his body.

The body as a fact

As we have seen, Sartre has secured the subject by positing an absolute self. We could suspect that this subject or self *inhabits* its body, but this would be idealism; on the contrary he declares that it coincides with it. However, wanting to avoid idealism, he only introduces a new form of it. Let us examine his position.

In the first place we must get rid of the illusion that my body can be an external object for me, consisting of a nervous system, glands, organs for digestion, for respiration, circulation, etc. The description of it is not about my body as I experience it, but about the body of nobody in particular, it is the description of the body as seen by biologists or doctors, it is the body of other people. *My* body for me is not an object in the middle of the world, an object to be studied. *It is for me much more a something I possess, than something I am.* This is a typically Christian conception: if it is our possession we are its masters.

A second and complementary illusion is that an absolutely objective approach is possible, because it could only be realised if the world would be a desert without humans, but such a world would not be the real world, as by the phenomenological definition of it, reality is the content of human consciousness. He then argues that science has admitted this impossibility to eliminate the subject. This contention is a strategy common to many metaphysicians: the world is necessary for the existence of men, and men are necessary for the existence of the world. Therefore, a Sirius point of view is impossible, as there is nobody on this celestial body to look at the universe.[47] In other words, an objective approach of the body leads to an abstract, bloodless representation that lacks the concreteness of the way an individual experiences it. My subjectivity is concrete in the sense that it is *somebody's* subjectivity; it is not the experience of the world and of my being present in it from *any* point of view whatsoever, but from *my* point of view and this point of view is determined by the *place* where I am, by the *situation* I am in. However, that I am somewhere, in some situation, is due to the fact that my body is *in the world*: I am here or there because I have a body. To be somewhere and to have a body is an *ontological necessity,* to be here is *contingent.* Thus, this necessity is also doubly contingent. In the first place it is not at all necessary that I exist; I do not exist in the mode of *causa sui*, I am not my own cause which would indeed make me necessary. In the second place it cannot be avoided that as a subject at any time I have a particular point of view, but that it is the point of view it is, is contingent. Or in his words: "In this sense one could define the body as *the contingent form taken by the necessity of my contingency.*"[48] This means that the only reason why a human being is not a universal ghost, without gender or age, without individuality, is that to be an individual is to have a body that determines its *human condition* and *situates* it in the world. This excludes its coincidence with the consciousness of ourselves or others; it excludes our universality and immortality.

The body "pour soi"

The body lends the consciousness finiteness: without body it would consist of an infinite possibility of contents. Indeed, because I have a body, my consciousness is not free-floating, but I am anchored in the world. I have a definite and therefore limited point of view that is my starting point, I make choices in certain situations, I decide that I have to be this or that, i.e. that I want to change things as they are. Thereby I perpetually transcend this body that I am *and am not*. Like Alice in *Through the Looking Glass*, human individuals must, according to Sartre, always run to stay in the same place:

> But from this point of view, which is perpetually overcome and perpetually reborn in the heart of this overcoming, this starting point that I do not cease to transgress and which is myself staying behind myself, is the necessity of my contingency.... I am my body to the extent that I am; I am not my body to the measure that I am not what I am; it is by my annihilation that I escape it. But I do not make an object of it for that reason, because I perpetually escape from what I am. And the body is necessary as an obstacle to surpass for being in the world, that is to say the obstacle I am to myself.[49]

In this sense the body is part of the world with its absolute order that I create and at the same time surpass towards what has yet to become, towards being beyond being.

My body for the other

Existentialist philosophers never consider individuals as unrelated atoms, they are with others and they cannot exist without the others. In Sartre's philosophy this being for others is central, and this relation is dialectical.[50] Where my body is concerned, it is known and used by others, but to the extent that I am a body for somebody else, this somebody else is revealed to me as that somebody for whom I am an object. As we have seen, we do not experience our own body as an object, the reason being that we coincide with it without being it and cannot step outside it. It is only in so far as others consider it as such and we are aware of this and reflect upon it, that we are capable of conceiving it as an object. Thus, if I go to the doctor because I am in pain, and the doctor examines me and explains me that I have a stomach ulcer, I become aware of my stomach as a stomach and of the ulcer in it. I begin to understand that my pain is an ulcer, comparable to those external ones I have seen before, but now in an organ that I cannot see, but that I now represent for myself as it is described to me, namely as having the form of a bagpipe. The body-object is the alienation of my pain; it is what I am without reason, without internal necessity, that what I cannot transcend towards something else. In summa: it is an obstacle for my freedom.

In his extensive analysis of the relation of individuals to one another, which he illustrates by referring to lovemaking, it becomes clear that like Plato he considers desire as an aim in se, alas doomed to end when incarnated and fulfilled.[51] This negative approach of the body and the central role accorded to subjectivity places him in the midst of Western tradition. Sartre considers sexuality as a fundamental, and perhaps *the* most fundamental relation between human beings, but he distinguishes between the conscious relation and the act. He stresses that the act in itself always signifies a failure, because, as said, the fulfilment is also the end of the desire for the other that is the central feature of lovemaking. Without the incarnation of the desire it could go on forever, but the coitus is not under the control of our consciousness. Indeed, the organs that are necessary for its accomplishment are commanded by the vegetative nervous system, not by our will:

> Anyway the desire is bound to fail. We have seen, indeed, that the coitus, which normally terminates it, is not its real aim. Certainly, more than one element of our sexual structure is the necessary translation of the nature of the desire, in particular the erection of the penis and of the clitoris. It is indeed nothing else but the affirmation of the flesh by the flesh. It is therefore absolutely necessary that it is not accomplished voluntarily; that is to say that we cannot use it like an instrument, but that it is, on the contrary, a biological and autonomous phenomenon of which the autonomous and involuntary accomplishment accompanies the sinking of consciousness into the body. What must be well understood is that no movable, prehensile organ tied to striated muscles could be a sexual organ, a *sex*; the sex if it had to appear as an organ could be nothing else but a manifestation of the vegetative life....[52]

From this we can conclude that Sartre was not, like Nietzsche a heathen, but in many ways related to the long Christian tradition of the aversion against the body and its instincts. This is confirmed by another text:

> ...We have simply wanted to stress that the sexual attitude is a behaviour directed to Somebody Else; that this behaviour contains necessarily the contingency of being-for-somebody-else and that of our own factuality, which goes without saying. But, that it is subjected from the start to a physiological and empirical constitution, that is what we cannot admit. From the moment "there is" a body and "there is" the *Other*, we react by *desire*, by *Love* and by the derived attitudes that we have mentioned. Our physiological structure only expresses symbolically and in the domain of the absolute contingency the permanent possibility that we are to take one or another of these attitudes. Thus we could say that the "Pour-soi" is

> sexual in its emergence before the Other and that, by the Other, sexuality comes into the world.[53]

The body in Sartre's conception remains *a necessary evil.* Desire and love are not caused by the body, but it is rather consciousness that is the source of sexuality.

When reduced to a thing, to flesh, our body is disgusting. In Sartre's opinion, to the extent that the body is objectively given, our will has no power over it; which is a scandal. Also, we cannot influence the way other people see it, not even clearly grasp the way they experience it, but only guess it. Rejecting dualism and idealism alike, phenomenalism is the solution: my body is the only object that I experience from within. I am this body in the sense that it permanently appears to me as the series of feelings, sensations, thoughts that constitute *me.* As we have seen, however, I never totally coincide with this stream of experiences: being *this*, I am already *something else*. In consequence, my ego cannot be pinpointed, because it is in fact always between two moments and therefore nothing. This explains why it is unacceptable for Sartre that the empirical body could be the ontological and epistemological basis for his anthropology. Though he rejects idealism of the Platonic type, he declares like Hegel that materiality as such has no *true* being and also that we are the only source of true being: "there is no other universe than a human universe, the universe of human subjectivity."[54]

His humanism

In *L'existentialisme est un humanisme* (Existentialism is a Humanism), he wants to clear up the many misunderstandings about his philosophy and to answer the criticism of his contemporaries. He tries to state under a simple and understandable form what his view on man is and what are its general implications. He wants to make clear that his conception of human existence is not to be confused with its fashionable interpretation and with the vague existentialist life style of the public frequenting the cafés and bars of Montmartre. It is not, he stresses, a pessimistic vision of man; it is neither quietism, nor a bourgeois philosophy as Marxists reproach him. His version of existentialism is atheistic and implies that the world is not made according to a plan and that man is not the exemplification of an essence. He distances himself of the philosophers of the Enlightenment, who deny the existence of the God of the Bible, but continue to believe in a world order and in an essence that precedes existence. The central thesis of his view is that the reality of human beings consists in leading their life without being defined by a pre-existing essence.

In *Saint Genêt, Comédien et Martyr* (Saint Genêt, Comedian and Martyr)[55], a work about the life of the writer Jean Genêt, he has developed this idea. Jean

Genêt was an orphan who was adopted. Like Sartre himself, the boy suffered from lacking a clear identity till the day his adoptive parents discover that he has stolen money and call him a thief. Now he knows what he is, he knows his essence: he is a thief and cannot be but a thief and this becomes his vocation. This is what we reduce ourselves to if we believe we are born with an essence, a once and for always given definite kind of personality. On the contrary, however, we are what we do, all what we do and only what we do. This requires that we make choices each moment of our life, and the only source and justification for these choices is our subjectivity: a man is what he decides to be, he is his own project.[56]

Sartre thus accepts the criticism that his philosophy is a subjectivism; more, he confirms it is its core. Indeed, a human individual cannot avoid making decisions, it is impossible not to, and he is the only one responsible for them. As we have seen, the subjectivity of an individual cannot be defined, because each of us is at each moment no longer what he was, yet the "ego" is the centre of action. This "ego" is immediately given and cannot be further analysed into its qualities. It is *a substance without properties*, and yet it is unique. Man has no nature, except absolute freedom, the only reason for which is that his choices are *his* choices, his projects are *his* projects. He cannot explain his decisions by his peculiar personality, by his rational judgement, by certain circumstances or influences, etc. However, let us not confound this freedom with André Gide's *acte gratuit*, which is a deed done for no reason at all, a caprice.[57] This absolute freedom is not a source of joy, but of anxiety, because the individual cannot shove off the burden of his responsibility upon other people, upon a supreme authority or upon a transcendent principle or God. What is even worse, the person by each of his acts commits all human beings, i.e. humanity itself. There is no justification for what he does other than that he chose to do it and that in a sense he cannot do wrong, because *he simply coincides with his choice.* Neither a priori in transcendent values, nor a posteriori, in e.g. a situation, can his decisions be given an external foundation. They are wholly due to his ego that, like Descartes' soul, cannot be located in time and space, but is without surface, a pinpoint upon which he has no grip. The only difference is that the ego of Sartre is not eternal and will not go to paradise.

Because men have no nature to refer to when making choices, the course of history cannot be predicted, and yet they cannot but act, and even if they do nothing at a certain moment this is a choice. However, the decisions of each and all of us determine what the future will be, and also what will be the face of humanity.[58] Each man is the sum of his deeds; his existence is a succession of moments that can be compared to the elements of an arithmetic series, the reason of which can only be discovered as it progresses. It is only at the end, when the last moment arrives, that a man knows for certain what his essence has been, which will also be constitutive for the essence of humanity. Thus, it makes no sense to say "Mankind is great!" referring to the merit of great individuals, or to their works. This would mean that there is a human nature after all, exemplified in Mozart and Goethe, in Einstein and Albert Schweitzer. Humanity does not yet exist, it can only be realised in the future, and this is what Sartre calls "nominalism," namely that an individual is nothing but the sum of countless deeds, humanity nothing but the sum of these sums when in the last minutes the addition is made.[59] For this reason, unlike Feuerbach and Comte, he does not declare that Man is the ultimate aim in the universe: Man must still be made.

Mauvaise Foi (Bad Faith) is the rejection of one's responsibility, seeking rationalisations for one's deeds. For example: "I did not become a great painter, because my parents did not allow me to study at the Academy of Fine Arts," or "I did not help this or that person, because I believed that someone else would come to his rescue." The "authentic" individual accepts that his acts are his own choice, which is the choice of his ego that he knows under its impure form as his empirical personality, but also as something absolute and unchanging that transcends the empirical data. Moreover this ego presupposes intersubjective relations, as indeed, we are not anybody, but somebody, a definite person, if we are recognised as such by other persons.

In conclusion

It is no wonder that Martin Heidegger in his *Brief über den "Humanismus"* (Letter about "Humanism") has explicitly distanced himself from Sartre's views. Heidegger criticises traditional Humanism, but not like Sartre because it presupposes a human nature, but because it is not a quest for Being, it does not look for the truth of man in the transcendent ground of all grounds. He, on the contrary, defines the humanity of man as the capacity to relate to Being.[60]

Marxists have severely criticised Sartre, because of his individualism. He distinguishes between the *human condition*, which replaces the old idea of human nature, and is universal and the *human situation,* which is unique for each individual. On the one hand, our common condition is that we are all mortal, that we must work to provide in our most essential needs and that we are social beings, a condition we cannot change. On the other hand, each of us is "thrown into the world," not only sharing the inevitable general aspects of existence, but also being born from particular parents, on a particular historical moment, in a particular society, which constitutes our primitive situation. We cannot alter our human condition, but our peculiar situation, though being a constraint upon our freedom, can be changed. By existing, we continuously surpass ourselves; according to Sartre: "each one of us makes the absolute by respiring" and "I construct the universe by choosing myself."[61] This is the core of the existentialist position, but it is problematic. It is based on Sartre's thesis that we *are nothing when we are born*, we are indeterminate and it is our subjectivity that will step by step constitute what we will be, by making project after project. Our situation, our genetic heritage, our physical condition, our environment and education, they all can be transcended. They are simply the cards we are going to play with, and perhaps we shall sometimes win, and perhaps we shall sometimes lose, but what matters is not winning or losing, but being free and responsible. The obstacles we find on our way are obstacles from the point of view of a definite project we have chosen. We can want to climb Mount Everest, and then this mountain is our obstacle, but only because we want to climb it. The essentialism of the Christian conception of man is rejected, but we can still recognise the original picture. Humans are unique by being subjects, persons; they have a free will and have a historical task in the world. Though there is no God to look "into our heart and kidneys," *if we are authentic we are perpetually observed by ourselves.* Each of us must scrutinise his conscience, the eye not turned to Heaven, but probing his own quality as a man, assessing his own existence as it has been hitherto, controlling his own authenticity. This is not the calm "know thyself" of Socrates, but a tragic pursuit, as indeed nobody ever can be certain to correspond to the high demands to which the individual is subjected in this philosophy. The deeds of each man must be exemplary; he must act as if the whole humanity was looking at him to follow his example, and this responsibility causes an acute anxiety. The effort resembles more that of the Christian, who, however hard he tries, though fervently wishing to follow in the steps of Jesus Christ, always falls short of his ideal. The difference is that the Sartrian existentialist must not answer to God. Yet, this equally obsessive introspection does not turn him into a solitary being, because he is in countless ways tied to other individuals. Taking an option through his own existence for the totality of humanity, he is solidary with the other beings sharing with him the human condition. We can consider this the Sartrian version of Christian love of one's fellow-man. Politically it is translated in his Marxism, with its vision of a society without classes at the end of times. Only when this is realised will man be totally human, freed of social and economical constraints and will he be able to become what he has chosen to become.

Sartre's Hymn to Freedom has been received enthusiastically by the youths of his time; they appreciated particularly his anti-bourgeois mentality, his progressive social and political views. Yet, it must be said that his picture of man remained idealistic and was typically Western. Starting from the human subjectivity he tries to make it as concrete as possible. This is done in two ways: *first*, we are in the world and we are in the world with others and thereby our subjectivity becomes and intersubjectivity, *second*, we are historical by being localised in time and space; together with others we are in a concrete political, economic and social situation. Defined as subjects condemned to absolute freedom, we are all equal, because our egos that make unique persons of us have no intrinsic qualities. Wanting our own freedom we want also the freedom of others.

According to Peter Sloterdijk, most idealist philosophers who preceded Sartre wanted to escape the loneliness of extreme subjectivism in a common reality, God (Fichte), Catholicism (Schlegel) or the State (Hegel) for example. Sartre accepted the abyssal freedom of the subject that is "an active nothingness," and enjoyed to "surpass" himself continuously. His moral dogma, however, was that he had to be solidary with all the oppressed people of whatever kind, because their freedom would be also his freedom. But to join forces with others was more a duty than a need. When he became a communist and engaged himself in the party, it was never wholeheartedly, but more an excuse for his genius and his uniqueness.[62] The main problem of his second important philosophical work after *L' Être et le Néant* (Being and Nothingness), *Critique de la Raison Dialectique* (Critique of Dialectical Reasoning),[63] was how to combine the unlimited freedom of the individual with the construction of a non-repressive society. His membership, his engagement and the tribulations with the leaders of the party that followed are of no direct interest for our subject, i.e. his philosophical point of view of man. All the details of his political position can be found i.a. in the extensive biography of Sartre by Annie-Cohen Solal.[64] The criticism I want to formulate has little to do with the difficult reconciliation of Sartre's individualism and the mentality of the communist partisan.

What Sartre can be reproached with is that the absolute freedom that according to him is the hallmark of humanity, remains completely theoretical. As history has proven, social situations, political situations, can in principle and in practice be altered, and individuals can assume their responsibility and change things. However, it is a "*statistical freedom*," not the freedom of each and every human being. Countless individuals don't have the capacity to make choices that are significant for the state of affairs in reality. People all over the world live in situations so extreme that it is virtually impossible for them to be anything else but the passive victims of a cruel fate. In the idealist view, man by his subjectivity is capable of directing the course of events, but here man himself is an abstraction. There is no such thing as "Man," "mankind," "humanity," these are mere names for all individual men. If subjectivity has to be the fundament of freedom, the thesis that constitutes the core of Sartre's anthropology, it has to remain abstract. In his theory men being born in different situations are not necessarily equal, but they are equally free. However, the moment their egos are seen as empirical, i.e. the result of their genetic properties, their external circumstances and their experiences, they have to be unique. If their egos have specific qualities, the freedom of the individuals to which they belong has to be specific too and freedom is relative instead of absolute.

Hegel said nature is full of exceptions and failures, because he compared the individual realisations with the ideal model. Even without such a comparison we know that many of us are ill-favoured by their genes, that they are born physically or mentally crippled. Others become incurably ill at some moment in their existence. It is of little comfort that they can cope with their illness in different ways and that this is still a choice. If the cards are too bad, and if before they are aware of it they become Alzheimer patients for example, the play is over. Proof is hereby given that we are no constant ego in a changing body, but that our mental capacities, our reason, our emotions, our will, coincide with the functioning of our brain. This means that right from the start our possibilities vary widely, and in consequence after all some are freer than others.

Schopenhauer was right that we are part of this world because we *are* bodies, instead of being subjects *transported* by them. Because of this fact we are subjected to the principles that govern nature and to the unpredictable course of history. In many situations individuals are powerless; their lives and those of their beloved ones are broken by dictatorial political regimes, war, famine, epidemics, not to speak of natural catastrophes. It is a lie and a scandal to say that we can always transcend our individual situation. Karl Jaspers, another existentialist philosopher, with the noblest intentions declared that even when we are threatened by a world catastrophe caused by an atomic bomb, we can protest by writing NO on a wall. Where are the walls where the victims of genocide in past and present could and can write NO? Where are the walls where the mother in the Sahel with her dying child in her lap can write NO? The Jew condemned to the gas chamber has no freedom but to die, and how he dies, crying and begging for mercy, or protesting, defying and insulting his executioners or accepting his fate, seems to matter little.

Sartre can be considered in the context of our concise history of the Western view of man to be a paradigm case. His anthropology, because this is what his existentialist philosophy is, is a continuation of a long tradition, first pagan with Pythagoras, Plato and Plotinus, then building further on it, becoming Christian with Saint-Augustine and the medieval Platonists, then taking new forms with Descartes' rationalism and Hegel's dialectical idealism. All along Man is considered a separate reality by his spirituality. He is a soul, a subject, an ego, of which the relation with the body is unclear. Different from the animals by his spiritual and/or rational capacities he is not really part of nature. For Western philosophers, whether Christian or atheist, all that matters is Man. Our world view is anthropocentric. If Nature is of interest, it is to the scientist, not to the philosopher or theologian.

Schopenhauer and Nietzsche were exceptions, who rejected the idealistic vision of man; they identified him with his body and reintegrated him in nature. No alternative came from the Anglo-Saxon countries: the utilitarian prevailed and was internalised to such a degree that a broader discussion of the fundamental nature of man "has given way to the discussion of means to assure the accepted end of mutual accommodation and individual discretion."[65] The challenges "Mother Earth" meets today are directly related to mankind, in the sense of human populations inhabiting it. Westerners have taken the lead and they export their visions, included their anthropocentrism. We can decide to cling to our most ancient beliefs concerning man, or to inform us about what science tells us.

We must be aware of the implications of an idealistic view of man: it has poor explanative value. It cannot explain why we are a successful species having colonized large parts of the world; why we continue to breed even under the poorest conditions when other species would stop to multiply; why we are an aggressive species, why we murder our congeners and why wars were and are waged. It cannot explain why the world economy grew to the point where we can no longer keep it in check and avoid its devastating effect on nature. We could continue the list of pressing questions to which science can give answers, even if they are provisional.

In the second volume of this book I intend to sketch the genealogy of the contemporary conceptions of Life and of our species proposed by scientists. Even if this alternative view of what it is to be human will certainly remain incomplete, it can contribute to a better understanding of ourselves and our relation with nature.

Notes

1. K. Löwith, "Nietzsches Versuch zur Wiedergewinnung der Welt," in: A. Guzoni, (ed.), *90 Jahre philosophische Nietzsche-Rezeption*, Hain, Meisenheim, 1979, 101.
2. N. Hartmann, *Der Aufbau der realen Welt*, W. De Gruyter & Co., Berlin, 1940, 195–197.
3. Hartmann, *Das Problem des geistigen Seins*, W. De Gruyter, Berlin, 1949, (1st ed. 1933), 55–56.
4. M. Scheler, *Die Stellung des Menschen im Kosmos*, in: *Späte Schriften*, S. Frings ed., Francke Verlag, Berlin, München, 1976, 44.
5. *Ibidem*, 67.
6. Cf. W. Stegmüller, *Hauptströmungen der Gegenwartsphilosophie*, Kröner, Stuttgart, 1965, 97.
7. Scheler, *Die Stellung des Menschen im Kosmos*, 31.
8. *Ibidem*.
9. *Ibidem*, 32.
10. *Ibidem*, 33.
11. *Ibidem*, 34. "With this process of becoming self-conscious, this new reaching back and focalising of his existence made possible by the Spirit, is given at the same time the *second* characteristic of the life of Man. By the power of his Mind, the creature we call 'Man' is not only capable to broaden what surrounds him to the dimension of World-being and to substantiate obstacles, but he is also capable to make—and that is what is most peculiar—*his own physical and psychical fabric and each of his own psychological experiences*, and to turn each of his vital functions into an *object* of his *consciousness*. Only for this reason is it possible for this creature to project freely his life."
12. *Ibidem*, 35.
13. *Ibidem*, 39.

14. *Ibidem*, 36.
15. *Ibidem*, 49.
16. *Ibidem*, 51.
17. *Ibidem*, 70.
18. *Ibidem*, 71.
19. Cf. Georges Minois, *L'Eglise et la Science*, Fayard, Paris, 1991, Vol. II, 245–248.
20. For a discussion of this problem see J. Rostand, *Hommes d'autrefois et hommes d'aujourd'hui*, Gallimard, Paris, 1966, 131 and following pages.
21. Cited by Minois, *L'Eglise et la Science*, II, 371, P. Teilhard de Chardin, "Hymne à la Matière," in: *Hymne de l'Univers*, Paris, 1961, 71.
22. Teilhard de Chardin, *Le Phénomène Humain*,152, note 1:

In fact, to the extent that it simply expresses the impossibility in which we are to perceive experimentally any living being (living or non-living) otherwise than engaged in a spatio-temporal series, evolutionism has already a long time ago ceased to be a hypothesis, in order to become a (dimensional) condition that must now be satisfied, in Physics and Biology, by all the hypotheses.—For the moment biologists and palaeontologists still debate about the modalities,—and before all about the mechanism of the transformations of Life: preponderance of (neo-Darwinian) Chance or (Lamarckian) play of invention in the appearance of new characteristics. But about the general and fundamental fact that there is as well organic evolution in the case of Life considered globally, as in that of whatever particular living being,—on this point, I say, all the specialists agree and for the good reason that they could not produce science if they thought otherwise... What can be regretted here (and not without astonishment) is that in spite of the clarity of the facts the unanimity goes not as far as to recognise (as admitted in these pages) that the 'galaxy' of the living forms traces a vast 'orthogenic' spiral movement towards more and more complexity and conscience."

23. Rostand, *Hommes d'autrefois et d'aujourd'hui*, 126–127. This is what Teilhard literally says:

So gradually are generally the ulterior modification along the phylum—also so stable sometimes, from the embryo on, of the organ (teeth, for example) that they affect, that we must definitely renounce to speak simply, in all the cases of the survival of the best adapted, or of mechanic adaptation to the environment and the use. Then what?

The more times I encounter and had to treat of this problem, the more the idea has imposed itself on my mind that in that case we witnessed an effect, not of external forces, but of psychology. Following our actual way of speaking, an animal would develop the instincts of a carnivore, *because* its molars are sharp and its paws possess claws. But must we not reverse the proposition? Said otherwise, if the Tiger has elongated its canines and sharpened its nails, would it not be precisely that, following his lineage it has received developed and transmitted 'the soul of a carnivore'? And would it not also be the case of the timid runners, the swimmers,—the burrowers—the gliders...? Cited from : Teilhard de Chardin, *Le Phénomène Humain*, Vol.1, I63–164.

24. Teilhard de Chardin, *Le Phénomène Humain*, 164.
25. *Ibidem*, note I, 208.
26. After an overview of different world cultures Teilhard de Chardin writes:

And it is thus that, step by step, we find ourselves drawn towards the more occidental zones of the World, — those, where, on the Euphrates, or the Nile, or the Mediterranean Sea, an exceptional encounter of places and peoples has taken place, in a few millenaries, and would produce a favourable mix thanks to which, without losing anything of their ascending force, on the contrary, reason would join the facts, and religion the action. Mesopotamia, Egypt, the Hellenic world— soon also Rome, — and above all this,... the mysterious ferment of Judeo-Christianity, lending its spiritual form to Europe!

...At this point of our investigation, one would have to falsify the facts, if one would not recognise that, during historical times, it is through the West that has passed the principal axis of the genesis of Man. In this zone of ardent growth and universal remoulding, all that makes our Man today has been found or at least *must have been re-found.* Because, even what had been known for a long time elsewhere, has only acquired its final human value by incorporating itself in the European system of ideas and activities. It is not simple naïveté to celebrate as a great event the discovery of America by Colombus.

...And proof of it is that invincibly, from one end of the world to the other, all the peoples, to remain human, or to become it more, come to pose, in the very terms in which the West has succeeded to formulate them, the hopes and problems of the modern Earth. *Le Phénomène Humain*, 234–235.

27. Teilhard de Chardin, *Le Phénomène Humain*, 306.
28. Rostand, Hommes d'autrefois et d'aujourd'hui, 140–141.
29. J. Passmore, *A Hundred Years of Philosophy*, Penguin, Harmondsworth, 1975, 494.
30. P. Foulquié, *Existentialisme*, 'Que Sais-je?,' Presses Universitaires de France, Paris, 1953, 67.
31. J.-P. Sartre, *Les Mots*, Gallimard, Paris, 1964, 66.
32. *Ibidem*, 39.
33. Sartre, *L'Etre et le Néant*, Gallimard, Paris, 1943, (1937 first edition), 290.
34. *Ibidem*, 293.
35. *Ibidem*, 301.
36. *Ibidem*, 20.
37. *Ibidem*, 34.
38. *Ibidem*, 58.
39. *Ibidem*, 59.
40. *Ibidem*, 65.
41. Sartre, *La Nausée*, Gallimard, Paris, 1938, 179
42. *Ibidem*, 181.
43. *Ibidem*, 223.
44. *Ibidem*, 224.
45. *Ibidem*, 181.
46. *Ibidem*.
47. Sartre, *L'Être et le Néant*, 365–370.
48. *Ibidem*, 371.
49. *Ibidem*, 391
50. *Ibidem*, 418–419.
51. *Ibidem*, 466
52. *Ibidem*, 466.
53. *Ibidem*, 477.
54. Sartre, *L'existentialisme est un humanisme*, Nagel, Paris, 1970, 93.
55. Sartre, *Saint Genêt. Comédien et Martyr*, Gallimard, Paris, 1952.
56. *Ibidem*, 23.
57. *Ibidem*, 74.

58. *Ibidem*, 53.
59. *Ibidem*, 58.
60. M. Heidegger, *Platons Lehre von der Wahrheit. Mit einem 'Brief über den Humanismus,'* Francke Verlag, Bern, 1947, 72:

Sartre formulates the fundamental thesis of existentialism as follows: existence precedes essence. He takes thereby existentia and essentia in the sense of metaphysics, which since Plato has been saying: essence precedes existence. Sartre reverses this proposition. But the reverse of a metaphysical proposition stays a metaphysical proposition. Thus he remains with traditional metaphysics in the oblivion of the Truth of Being.

61. Sartre, *L'existentialisme est un humanisme*, 68.
62. Peter Sloterdijk, "Vorbemerkung" in: Peter Sloterdijk, (ed.), *Sartre*, Deutscher Taschenbuch Verlag, 1998, 8–9
63. Sartre, *Critique de la Raison Dialectique*, Gallimard Paris, 1960.
64. A. Cohen-Solal, *Sartre* ,1905–1980, Gallimard, Paris, 1985.
65. *The Encyclopaedia of Philosophy*, Paul Edwards, editor in Chief, Macmillan Publishing and Free Press, New York, London, 1967, Vol. V.

Bibliography

Abendroth, W., *Schopenhauer*, Rohwolt, Hamburg, 1967.

Albertz, R., *Religionsgeschichte Israels*, Vandenhoeck-Ruprechts, 1992, transl.: *A History of Israelite Religion in the Old Testament Period*, 2 vols., SCM Press, 1994.

Aristotle, 23 vols., Loeb Classical Library, Harvard Univ. Press, Cambridge Mass., 1995.

Bloch, E., *Über Karl Marx*, Suhrkamp Verlag, Frankfurt am Main, 1980.

Bowie, F., *The Anthropology of Religion*, Blackwell, Oxford, 2002.

Boehner, Ph., O.F.M., *Ockham, Philosophical Writings*, Hackett, Indianapolis, 1990.

Bréhier, E., *Histoire de la Philosophie*, 8 vols., Presses Universitaires de France, Paris, 1951.

Campbell, J., *The Masks of God,* 5 vols., Penguin, Arkana, Harmondsworth, 1991.

Clapp, B.W., *An Environmental History of Britain since the Industrial Revolution*, London, New York, 1994.

Cohen-Solal, A., *Sartre ,1905-1980*, Gallimard, Paris, 1985.

Damasio, A. R., *Descartes' Error. Emotion, Reason and the Human Brain*, Papermac, London, 1996.

Dawkins, R., *The God Delusion*, Black Swan, London, 2007.

Descartes. Oeuvres, La Pléiade, Gallimard, Paris, 1953.

Evans, G. R., *Philosophy and Theology in the Middle Ages*, Routledge, London, 1993.

Foulquié, P., *Existentialisme*, "Que Sais-je?," Presses Universitaires de France, Paris, 1953.

Farrington, B., *Greek Science. Its Meaning for Us*, Penguin Books, Harmondsworth, 1944.

Gibson, J.J., *The Ecological Approach to Visual Perception*, Houghton Mifflin Co., Boston, London, 1979.

Gilson, E., *Le Thomisme*, Librairie Philosophique, J. Vrin, Paris, 1948.

Gilson, E., *La Philosophie au Moyen-Age*, Payot, Paris, 1962.

Gombrich, E.H., *The History of Art*, Phaidon, 1986.

Gosling, J.C.B., *Plato*, Routledge and Kegan Paul, London, 1983.

Guthrie, W.K.C., *The Greek Philosophers. From Thales to Aristotle*, Routledge, London, New York, 1989.

Gosselin, M., "Quine's Hypothetical Theory of Language Learning. A Comparison of different conceptual schemes and their logic," in: L. Decock, L. Horsten (eds.), *Quine. Naturalized Epistemology, Perceptual Knowledge and Ontology*, Rodopi, Amsterdam, Atlanta, GA 2000.

Guzzoni, A., *90 Jahre Nietzsche Rezeption*, Hain, Königstein, 1979.

Hamelin, O., *Le Système d'Aristote*, edited by L. Robin, Félix Alcan, Paris, 1931.
Hartmann, N., *Der Aufbau der Realen Welt*, W. De Gruyter & Co., Berlin, 1940.
Hartmann, O., *Das Problem des geistigen Seins*, (1rst edition 1933), W. De Gruyter, Berlin, 1949.
Hegel, G. W. F., *Phenomenologie des Geistes*, Felix Meiner, Hamburg, 1952.
Hegel, G. W. F., *Die Vernunft in der Geschichte*, Felix Meiner, Hamburg, 1955.
Hegel, G. W. F., *Enzyklopädie der philosophischen Wissenschaften*, Felix Meiner, Hamburg, 1959.
Hegel, G. W. F., *Wissenschaft der Logik*, Felix Meiner, Hamburg,1963.
Heer, F., *Hegel*, Fischer Verlag, Frankfurt am Main/Hamburg,1955.
Heidegger, M., *Platons Lehre von der Wahrheit. Mit einem "Brief über den Humanismus,"* Francke Verlag, Bern, 1947.
Hoffmeister, J., "Einleitung des Herausgebers," in: *Phänomenologie des Geistes*, Meiner, Hamburg, 1952.
Hume, D., *A Treatise of Human Nature*, 2 vols., Everyman's Library, London, 1964.
Hyman, H., Walsh, A., (eds.), *Philosophy in the Middle Ages*, Hackett, Indianapolis, 1973.
Janz, Curt Paul, *Friedrich Nietzsche Biographie*, Hanser, München, 1978.
Koestler, A., *The Sleepwalkers*, Penguin, Pelican, Harmondsworth, 1986.
Köller, G., Anders, F., *Zoölogie*, Carl Winter-Universitätsverlag, Heidelberg, 1963.
Kreglinger, R., *La Religion d'Israel*, M. Lambertin- E. Nourry, Bruxelles/Paris, 1926.
Leff, G., *Medieval Thought. (St Augustine to Ockham)*, Penguin, Harmondsworth, 1961.
Leff, G., *Wiliam of Ockham*, Manchester University Press, Manchester, 1975.
Lévi-Strauss, Cl., *Tristes Tropiques*, Plon, Paris, 1955.
Lévi-Strauss, CL., *Race et Histoire,* Editions Gonthier, Paris, © Unesco, 1961.
Lévi-Strauss, *La Pensée Sauvage*, Plon, Paris, 1962.
Locke, J., *Essay concerning Human Understanding*, 2 vols., Everyman's Library, Dutton, New York, Dent, London, 1967.
Lorenz, K., "The Role of Gestalt Perception in Animal and Human Behaviour," in: *Aspects of Form*, Lancelot Law Whyte, (ed.), Lund Humphries, London, 1968.
Löwith, K., *Meaning in History*, Phoenix Books, University of Chicago Press,1949.
Löwith, K., *Nietzsches Philosophie der ewigen Wiederkehr des Gleichen,* Kohlhammer, Stuttgart, 1956.
Löwith, K., "Nietzsches Versuch der Wiedergewinung der Welt" in: A. Guzzoni (ed.), *90 Jahre Nietzsche Rezeption*, Hain, Königstein, 1979.
Minois, G., *L'Eglise et la Science*, 2 vols., Fayard, Paris, 1991.

Mack, B. L., *The Lost Gospel. The Book of Q and Christian Origins*, Element, Shaftesbury, Dorset, Rockport, Massachusetts, 1993.
Marx, K., *Deutsche Ideologie*, in: *Die Frühschriften*, A. Kröner, Stuttgart, 1963.
Marx, K., *Das Kapital*, 3 vols., Dietz Verlag, Berlin, 1962.
Morris, D., *The Naked Ape*, Jonathan Cape, London, 1967.
Mole, P., *Acupuncture*, Element Books, Shaftesbury, 1992.
Plato. The Collected Dialogues, Bollingen Series LXXI, Princeton University Press, Princeton, 1999.
Pagels, E., *The Gnostic Gospels*, Random, Vintage, New York, 1981.
Passmore, J., *A Hundred Years of Philosophy,* Penguin, Harmondsworth, 1975, (first edition 1957).
Popper, K., *The Open Society and its Enemies*, 3 vols., Routledge, London, 1945.
Rostand, J., *Hommes d'autrefois et hommes d'aujourd'hui*, Gallimard, Paris, 1966.
Russell, B., *A History of Western Philosophy*, Allan and Unwin, London, 1946.
Safranski, R., *Schopenhauer und die wilden Jahre der Philosophie*, Carl Hanser Verlag , München, 1987.
Sartre, J.-P., *La Nausée*, Gallimard, Paris, 1938.
Sartre, J.-P., *L'Etre et le Néant*, Gallimard, Paris, 1943, (1937 first edition).
Sartre, J.-P., *Saint Genêt. Comédien et Martyr*, Gallimard, Paris, 1952.
Sartre, J.-P., *Critique de la Raison Dialectique*, Gallimard Paris, 1960.
Sartre, J.-P., *Les Mots*, Gallimard, Paris, 1964.
Sartre, J.-P., *L éxistentialisme est un humanisme*, Nagel, Paris, 1970.
Schalk, F. "Die europäische Aufklärung" in: *Propyläen Weltgeschichte*, vol.7, Golo Man, August Nitschke, (eds.), Propyläyen Verlag, Berlin/Frankfurt am Main, 1986.
Scheler, M., *Die Stellung des Menschen im Kosmos,,* in: *Späte Schriften*, (ed. S.Frings), Francke Verlag, Berlin, München, 1976.
Schlechta, K., *Nietzsche Werke*, 3 vols., Hanser, München, 1956.
Schlechta, K., *Nietzsche Index*, Hanser, München, 1965.
Schlechta, K., *Nietzsche Kronik. Daten zu Leben und Werk*, Hanser Verlag, München, 1975.
Schopenhauer. Werke, 5 vols., Wissenschaftliche Buchgesellschaft, Darmstadt, 1982.
Sloterdijk, P., "Vorbemerkung" in: Sloterdijk, P., (ed.), *Sartre,* Deutscher Taschenbuch Verlag, München, 1998.
Stegmüller, W., *Hauptsrömungen der Gegenwartsphilosophie*, Kröner, Stuttgart, 1965,
Taton, R., *La science antique et médiévale. Des origines à 1450*, Quadrige, Presses Universitaires de France, 1994.
Teilhard de Chardin P, *Oeuvres de Pierre Teilhard de Chardin,* 6 vols., Editions du Seuil, Paris, 1960-1962.

The Encyclopaedia of Philosophy, 5 vols., Paul Edwards, (ed.), Macmillan Publishing and Free Press, New York, London, 1967.

The Revised English Bible, (RVE), Cambridge University Press, Cambrridge, 1989.

The Fontana Dictionary of Modern Thought, Bullock, A., Stallybrass, O., Trombley, S. (eds.), Fontana Press, London, 1988.

Tricot, J., (introduction, notes and translation), *Aristote, La Métaphysique*, Vrin, Paris, 1970, 2 vols.

Wallerstein, I., *The Modern World-system, Capitalist Agriculture and the Origins of the European World-economy in the Sixteenth Century*, Academic Press, New York, London, 1974.

Weinberg, J.R., *Abstraction, Relation and Induction*, University of Wisconsin Press, Madison & Milwaukee, 1965.

Index of Names

www.ingramcontent.com/pod-product-compliance
Ingram Content Group UK Ltd.
Pitfield, Milton Keynes, MK11 3LW, UK
UKHW020126290926
15471UKWH00003B/311

9 780761 865193